# FIREWORKS

# Principles and Practice

by

**The Reverend Ronald Lancaster M.A.**
Chaplain of Kimbolton School, Huntingdon, England
Firework Consultant to Pains-Wessex Ltd, Salisbury, England.

and

**Takeo Shimizu, Doctor of Engineering.**
Director of the Koa Firework Co., Tokyo

**Roy E.A. Butler M.A.**
a Master at Kimbolton School

**Ronald G. Hall F.R.I.C.**
Technical Director, Brock's Crystal Palace Fireworks Ltd.
Hemel Hempstead, England.

CHEMICAL PUBLISHING CO., INC.     New York  1972

# Fireworks, Principles and Practice

ISBN: 978-0-8206-0216-5

Chemical Publishing Company:
www.chemical-publishing.com
www.chemicalpublishing.net

First edition:
© **Chemical Publishing Company, Inc.** – New York, 1972
Second Impression:
**Chemical Publishing Company, Inc.** - 2011


Printed in the United States of America

# Preface

For many years Weingart's "Pyrotechnics" has been regarded as the amateur firework enthusiast's Bible, and it was news of the re-print of this work in 1968 which prompted the writer to suggest a revision of it. As it happened the suggestion came too late with the result that a new work has evolved.

From the beginning the writer was anxious to share the task of writing this work, and accords grateful thanks to the other three contributors:- Dr. Shimizu, who very willingly translated part of his book "Hanabi" from the original Japanese. The script of chapter 19 is more or less as he translated it, and a great credit to him. To the best of our knowledge this is the only treatise on Japanese firework manufacture in the English language.

Ronald Hall, one of my long-standing firework friends who has long experience as a chemist in the explosives and firework industries. Has also been responsible for the introduction of polymerizing resins into commercial firework manufacture and is especially interested in forensic aspects of explosives.

Last but not least my thanks go to my teaching colleague and friend Roy Butler; an able firework maker who has given even more of his time to write a precis of the available historical records, adding also more up-to-date material.

Turning to the general preparation of the book, I would like to expresss grateful thanks to Peter Smout Esq., M.A., Senior Master at Kimbolton School who has so kindly read through the script and made many helpful suggestions.

Helpful comments have also been made by Peter Watson Esq., B. Sc. Senior Chemistry Master at Kimbolton School, Dr. Herbert Ellern, the author of Military and Civilian Pyrotechnics, and Mr. J. Barkley and Mr. J. Wommack, two other American friends. My wife, Kathleen Lancaster, Dip. Arts, Dip. Ed., has kindly assisted with drawings and diagrams along with P. R. Lambert, a member of the School Sixth Form.

In particular also my grateful thanks go to Edwin Bailey who kindly used his printer's expertise to convert many of the drawings into a suitable form for printing.

Several commercial firms have been kind enough to supply technical information. These were Imperial Chemical Industries, Albright and Wilson Ltd., Frederick Allen & Sons Ltd., Anchor Chemical Co. Ltd., F. W. Berk & Co., Ltd., Columbian International Ltd., Du Pont de Nemours & Co., K. W. Chemicals Ltd., W. S. Lloyd Ltd., Magnesium Elektron Ltd., Chas. Page & Co. Ltd., L. R. B. Pearce Ltd., A. F. Suter & Co. Ltd. and Bush Beach, Segner Bayley. I would like to express my gratitude to all those people who helped me along the firework road in those early days when help was required to cross the threshold which separates amateur and professional firework manufacture. In particular I would mention the Greenhalgh Family of Standard Fireworks Ltd., Huddersfield, along with W. Stott Esq. and J. Seymour Esq. who also live in Huddersfield, my native town. Kindly friends abroad include Lünig of Stuttgart; Nico of Trittau, Hamburg; Moog of Wuppertal; Hamberger of Oberried and the Barfod Family of the Tivoli Gardens in Denmark.

Lastly, and in more recent times, gratitude is due to Pains-Wessex Ltd. to whom I have been Firework Consultant since 1963 and to John Deeker F.C.A. and David A.S. Little for their help and friendship.

Ronald Lancaster
7, High Street,
Kimbolton
Huntingdon
England.

# Introduction

It is illegal to manufacture fireworks in most countries unless a license has been obtained from the government. This is absolutely right, for nowhere else does the old saying that "a little knowledge is a dangerous thing" apply more than here, perhaps with disastrous effects. Accidents occasionally happen in the most experienced hands and old and hardened manufacturers shudder at some of the experiments of the uninitiated.

Why then write a book about fireworks?

There is a need for an up-to-date description of *general* firework practice. Firework manufacture may be a mixture of chemistry and cooking, but it is nevertheless an important branch of pyrotechnics. All the books in existence lack either accurate detailed information or publish information which may be incorrect, dangerous or useless commercially. Naturally this has been deliberate because firework manufacture has been in the hands of private families and is still more or less entirely tied up with money and competition. This is a pity, but like so many commercial enterprises, considerable sums of money are invested in plants or research and returns are naturally expected. Indeed, the writer has done little more than skim over the surface, quite deliberately; nevertheless all the compositions are typical of those in use in Europe and are as reasonably safe as such things can be. Clearly the intention of this book has been to attempt to show that much of the available printed information is dangerous.

Over the last few decades the attitudes of the manufacturers have changed. In the past each one regarded his compositions as a great secret, the "boss" himself frequently doing the mixing and giving the chemicals false names to fool the industrial spies. All this has more or less gone. Chemical suppliers became fewer and larger, selling the same materials to everyone; gunpowder manufacture is virtually a monopoly, and fireworkers in some countries change their employment from one company to another.

Most good firework makers share the same basic formulations; only the finer points and the techniques are more or less secret and naturally these are details which do not reach publication. In any case half the battle of firework manufacture is experience, namely the constant observation of the burning characteristics and performance of fireworks, and consequently the experience of knowing what adjustments to make and what to look for.

In the opinion of the writer, the argument that explosive information should not be published, does not hold water. Determined people can get a good deal of information, for there is plenty of it in print, and after all, legal and other restrictions make it very difficult for anyone to start manufacture.

The writers naturally would be greatly disturbed to feel that this book has caused anyone to damage people or property but such risks have to be taken at all levels of life. Fireworks are dangerous but so are domestic electricity supplies, oil burning heaters, pans of boiling fat, gasoline pumps, gas supplies, children's bicycles on roads......the possibilities are endless.

From time to time attempts are made to ban the sale of fireworks to the public. Recent voting in Great Britain indicated that the majority of the voters were against such a move, and quite rightly so. After all people have to act responsibly and should be free to exercise their responsibility in this direction. Britain, in common with most European countries, has rigid legislation and inspection of firework manufacture and an agreement amongst manufacturers that flash crackers and certain dangerous fireworks should not be sold to the general public. The result is that a fairly wide range of fireworks can be purchased in the shops at certain times of the year, and display fireworks can be organized by people with specialized experience. The U.S.A. could do well to benefit from our experience, for it would appear that a country priding itself on its freedom can nevertheless allow some bureaucratic fire marshall or other excited group to bring in legislation to outlaw fireworks in individual states. The result appears to be that it encourages people to buy fireworks over the border in a more permissive state and fire them illegally. Restrict the dangerous explosive items by all means, but "safe and sane" as the Americans put it, covers *very* much more than sparklers.

The Germans say in effect that once a person has smelt blackpowder, he will be with it for the rest of his life. There is undoubtedly some truth in this, for real fireworkers all over the world love to get together and talk about the fascination of this, their mutual interest. It is to be hoped that it will always be possible to strike a happy balance between the enthusiast and the legislation.

In recent years while pyrotechnics have been striding ahead, the art of firework manufacture appears to be relatively static and oldfashioned. Nevertheless this should not be a matter exciting too great a concern, for the firework maker can only display his art on those grand and comparatively rare occasions when large sums of money are spent on a single display. The burst of an 8″ golden octopus, crossette shell or a Japanese chrysanthemum will still thrill people for many years to come, in spite of the fact that the composition may be primitive. Public taste will not have the opportunity to become bored by those fireworks which really display the firework maker's art.

# Contents

# Chapter 1

# The History of Fireworks

## R. E. A. Butler

Firework manufacture has a long history, but the development of the pyrotechnic art has been remarkably slow. The Chinese may have made fireworks of sorts over a thousand years ago; displays have been fired at public and private celebrations for five hundred years, and their popularity, now throughout the world, seems undiminished. Nevertheless, basically firework displays have changed little over several centuries, and rockets, shells and Roman candles, in various forms, remain the main display components. Certainly colors have been improved, and the range of colors extended, shells are more spectacular, rockets are propelled higher, the use of new materials has brought some new effects, and set pieces and the style of displays have been changed, but the essential ingredients of the firework exhibition do not alter. The fireworker still strives to excite and delight with a combination of color and noise. He creates patterns of beauty and brilliance using natural materials and employing a knowledge of chemical reaction, together with the benefits of experience, and often much patience, dedication and intuition. The beginning of the pyrotechnic art was heralded by the invention of gunpowder, and this dark mixture is still the firework maker's principal material. Thus, in this capacity as a bringer of pleasure and beauty, gunpowder makes some amends for its evil reputation as a source of death and destruction.

It is probable that the first gunpowder was formed when, quite by chance, charcoal, saltpeter and sulphur were brought together. The result of this accident would be obvious if the mixture was exposed to some means of ignition, and the potential use of this new explosive material would soon become apparent. Traditionally the Chinese are credited with the discovery at a time well before historical records. Certainly the evidence suggests that gunpowder originated in the East,

1

with China or India being the likely source, although the Arabs and Greeks have certain claims. Moreover, coded writings by the English friar, Roger Bacon, in the thirteenth century, are generally accepted as a description of a gunpowder mixture for the production of an explosion. The invention of the gun, which probably represented the greatest step forward in the application of gunpowder, was almost certainly made in Germany, at Freiburg, by a Franciscan monk called Berthold Schwarz, although the inventor could well have had Asiatic origins. It was over two centuries later that the first artillery was seen in China, and that was on Portuguese ships in 1520.

However, the Chinese had employed pyrotechnic mixtures long before this date. Ancient manuscripts describe explosive bombs, which were fired from giant catapults, and burst on landing or in the air. Similar missiles were merely dropped on the enemy from fortress walls. Firecrackers were used in early times, just as they are now, to scare away evil spirits from wedding and birth celebrations, and from funerals, and they were also much in evidence at various religious festivals. These crackers were often made by packing gunpowder into bamboo cases or rolled paper tubes, so laying the foundations of modern firework making.

An encyclopaedia by Fang I Chih in about 1630 included a mention of 'fire trees and silver trees' used in the Tang dynasty (7th to 10th centuries) in which gunpowder was thought by the author to have been used. These fireworks may have been forerunners of those used in big displays which were frequently put on in China in the seventeenth and eighteenth centuries, and which were described in various writings by travellers returning to Europe. Apparently the development of Chinese fireworks proceeded very slowly, and in 1821 Claude-Fortune Ruggieri, the French pyrotechnist, remarked that his information was that the 'Chinese fireworks ............ were no different from what the Chinese have been making for three or four centuries; this convinced me that we in Europe are far superior to the Chinese'.

In India too progress appears to have been slow, for war rockets were in use at a very early time. Here, as in China, fireworks of sorts were frequently seen at celebrations and public festivities, and fifteenth and sixteenth century writings, such as the Marathi poem of Saint Ekanatha, describe displays, and mention rockets and fireworks producing garlands of flowers, a moonlight effect and hissing noises. By the eighteenth century displays were being organized on a lavish scale.

The first English display in India was in 1790 near Lucknow, and was said to have taken six months to prepare.

In Europe pyrotechnics for military purposes saw an early peak of achievement in the form of Greek fire. Highly combustible material, including sulphur, resin, camphor and pitch, was blown by a bellows device out of copper or iron tubes, or even handpumps, and was almost inextinguishable. Old manuscripts suggest several ways of attempting to combat the fire, especially the application of wine, vinegar, sand and even urine. For four hundred years the Greeks guarded the secret of their devastating weapon, and used it with spectacular effect on land and sea; but by the tenth century the Saracens had learned the formula, and used it against the Crusaders. By the fourteenth century gunpowder had made its appearance in European warfare, and made the short-ranged Greek fire powerless against far-flung missiles.

In the wake of gunpowder came the arrival of firework mixtures, both of them appearing in Europe, probably as a result of information on their manufacture being brought from the East. Italy seems to have been the first area in Europe to make fireworks, as opposed to military pyrotechnics, and to put on displays. It is clear that before 1500 fireworks were employed extensively at religious festivals and public events, and frequent displays were becoming popular entertainments. Florence was probably the center of an expanding manufacturing industry, as demand for the new spectacle increased. Before this period fireworks had been used as scenic effects at theatrical productions. In fact, fiery torches and the like had been added embellishments in the amphitheatres of classical Roman times. Now the fireworks became the main concern, although elaborate scenic sets and buildings were to form backgrounds to displays for many years to come.

Firework displays were seldom seen in England before the end of the sixteenth century. Shakespeare refers to 'fireworks' on several occasions in his plays, suggesting that the term was in general usage in England at that time. Other literature of the period often mentions the 'green man' whose function was to walk at the head of processions carrying 'fire clubs' and scattering 'fireworks' (in this case probably meaning sparks) to clear the way. The origin of this character and his title are a mystery, but we are told that he was usually made up to appear very ugly, and he certainly survived well into the next century.

The earliest record of a firework display in England was in 1572, when a large show was put on at Warwick Castle to mark the visit of Queen Elizabeth I. The Queen is said to have enjoyed the spectacle immensely, and this approval served to encourage the organization of many more displays, including two shows fired at Kenilworth Castle, Warwickshire, to entertain Her Majesty during a visit there in 1575. The first of the displays on the River Thames was in 1613 to celebrate the marriage of King James's daughter, Elizabeth. The site has been used with great regularity ever since.

The early displays in England were mainly the work of firework makers from France and Italy, especially the latter, who seem to have been supreme in Europe until the end of the seventeenth century. It was not until considerably later that English pyrotechnists began to challenge the continental lead. Responsibility for the provision of fireworks and the organization of displays was put in the hands of the military, and Ordnance officers, ranked Firemasters, were appointed to take charge.

While the English lagged behind, two distinct schools of firework making became apparent in Europe. In the Northern area, such states as Poland, Sweden, Denmark and the German states were developing new methods of firework presentation, which differed markedly from the traditional style of the Mediterranean countries. Brock considers

Fig. 1  A more modern type of "Machine" (Courtesy J. Wommack, Esq.)

that the split was closely related to religious matters, and that the intense feelings which the Reformation aroused found outlet in more sectarian spheres, including pyrotechnics. In fact, the fireworks made in the north and south remained very similar in effects; the divergence was more in the composition of displays.

The Italian style, illustrated especially by the Ruggieri brothers of Bologna, and followed by the manufacturers of France (who were joined by the Ruggieri family at a later date), had grown from the early ceremonial displays on saints' days and religious festivals. Among these were the annual displays in Florence at the Feast of Saint John and the Assumption, and those at Rome to mark the Feast of Saint Peter and Saint Paul. Invariably collections of small fireworks were arranged on, and in front of, huge, elaborate structures, built in the form of castles, temples or classical edifices, and known as 'machines' or 'temples'. The imposing frontages were lavishly adorned with rich decorations, and the whole was illuminated from without and within. The audience was thus entertained before the actual display began, and when the fireworks were lit they tended to heighten the general spectacle of the 'machine', rather than provide purely pyrotechnic amusement. A more modern machine is shown in Fig. 1.

The breakaway Northern school took their lead from Nürnberg, where experts like Hoch, Muller, Clarmer and Miller, challenged the masters of Florence and Bologna. The displays in the North gave the actual fireworks the prime importance and diminished the role of the 'machine'. The fireworks were set out in neat rows on the ground for all to inspect before the display was fired. If a 'machine' was used, it was a much less elaborate construction than in the South, although sometimes real buildings or landscapes were utilized to add atmosphere to the shows. The effect achieved by firing displays behind a foreground of water was realized in this period, and engravings of displays at Stockholm, Paris, Versailles and on the Thames, illustrate the early beginnings of this still popular practice.

Spectators accustomed to either the Northern or Southern type of display were scornful of the attempts of the rival school, as can be clearly discerned from contemporary publications. The most authoritative was 'The Great Art of Artillery', penned by Casimir Siemienowitz in 1650. His displays, although following the techniques of the Northern school in the main, included some features from the South, so giving his shows decorative effect before firing time, yet concentrating on pure firework amusement during the performance. This kind of compromise display often included figures and architectural structures, smaller and less intricate than the 'machine', and made of a wooden frame, over which was papier-mâché, which concealed fireworks. At a certain point in the show sparks and stars would be seen to issue from the model with spectacular effect. Various figures made their appearance in the different shows, although the Cupid was perhaps the most popular, and the tall 'obelisk' was a regular feature at displays for many decades.

In their various styles displays increased enormously in number and size all over Europe. Louis XIV and XV enjoyed numerous shows in Paris in celebration of royal birthdays and weddings, state occasions and victory or peace festivals. In almost every European country visiting royalty were invariably treated to displays. Peace treaties, like that signed at Aix la Chapelle in 1742, were excuses for expensive performances in many European capitals. Numerous prints and engravings of the time undoubtedly flatter many of these shows by depicting them always in full, extravagant splendor. In fact, not all the displays were the spectacular success their advance publicity proclaimed.

The pyrotechnic celebration planned to take place in London's Green Park in 1742 was to have been the greatest display of all time. An official estimate of the cost was over £14,500, and Ruggieri and other notable Italian manufacturers were brought over especially for the occasion. Nearly six months were spent in erecting huge 'temples' and various ornate 'machines' of elaborate design. The organizers even engaged Handel to compose a musical work in honor of the event, and the popular 'Fireworks Music' was the result.

At the appointed hour, King George II, accompanied by an impressive array of aristocracy, paraded to his seat past the huge, excited crowd. However, all was not well behind the scenes, for violent arguments had arisen between the English and Italian fireworkers. These disagreements were brought to a dramatic end as an explosion rent the North Pavilion, which burst into flames. The fire caused widespread confusion and alarm, but was eventually brought under control so that the planned fireworks could begin. However, judging by eyewitness reports, the display was anything but the memorable spectacle which had been promised. Such descriptions as 'pitiful and ill-conducted', 'the Grand Whim for posterity to laugh at' and 'the machine was very beautiful and was all that was worth seeing' were just some of the less abusive comments. Certainly it was the last big display London was to see for many years.

Private firework companies had for long been operating on the continent of Europe, but in England artillery officers were still in charge of displays, although the actual arrangements were probably under the control of civilians. No doubt small English companies made fireworks for the shows, and large quantities were regularly imported from France and Italy. It is recorded that a Swede, Martin Beckman, made the fireworks for the celebrations which marked the coronations of Charles II and James II, and also that of William of Orange. However, the eighteenth century, 'the Age of Elegance', gave the English manufacturers the opportunity to show their skills and to increase their sales and production.

It was during this period that the 'Pleasure Garden' became, for the respectable townsmen and their ladies, the fashionable place at which to be seen. Taking a lead from London, most towns of note established those exclusive resorts with their concerts and tea parties, opportunities to exchange gossip and to be sociable, not to mention the availability of medicinal waters in many parts of the country. Soon other entertainments were added to amuse and excite clientele. Male and female bare knuckle fights, dog fights and bear and bull baiting were all popular attractions, and eventually firework displays became regular items on the programs. Many small manufacturers found this new and expanding market just the incentive they needed to develop their businesses and make their reputations. The Brock family business, in particular, made great strides during this period, with impressive displays at the famous Marylebone Gardens, and later at Ranelagh, Vauxhall and the Spa Gardens, Bermondsey. London boasted scores of resorts, ranging from the most fashionable and exclusive, to others which were rather less notable and often short-lived. Outside the capital most towns of any size had their own Gardens. The most outstanding of these were the Belle Vue Gardens in Manchester, which still exist as a sporting and amusement centre, a zoological garden and a site for regular firework displays. Other long-established parks which offer fireworks as major attractions in the summer months include the Tivoli Gardens of Copenhagen and the Lisseberg at Gothenberg.

The vast and rapid expansion of building on the outskirts of London and other towns during the early part of the nineteenth century inevitably caused the demise of most of the pleasure resorts, although some of them were converted into public houses and still serve, if for a wider clientele. The firework makers were obliged to turn their main efforts towards the production of large and spectacular displays which celebrated events of national importance. Notable among these were the exhibitions which marked the Peace Treaty of 1814, the Jubilee of George III, and the coronations of George IV in 1821 and Victoria in 1838. These expensive Hyde Park displays rivalled in effect the many held in Paris at this time, especially at the request of Napoleon, while he ruled France, since he was a great firework enthusiast and had much to celebrate before his ultimate defeat.

As the nineteenth century wore on, a dramatic leap forward was seen in the techniques of firework making, a progress which mirrored the vast development which was proceeding in all fields of scientific research and means of communication in Europe. In time, the better understanding of chemical reactions produced new pyrotechnic effects and especially a wider range and a greater intensity of colors. A better use of propellants and more efficient methods of filling fireworks led to ever improving pieces and consequently better displays. Newly discovered knowledge was circulated by the numerous scientific books which were being published, encouraging a large and new generation of amateur pyrotechnists. The expanding sales of newspapers and journals like 'The Illustrated London Times' and 'The Illustrated News of the World' publicized and popularized displays with frequent and large illustrations. Better means of transport enabled manufacturers to put on displays in areas which had previously been regarded as inaccessible, and allowed people from increasingly wider circles to travel to watch the exhibitions. Moreover, all over Europe, Royalty and other dignitaries, were able to move about the Continent with much greater facility and more often than not were treated to a spectacular firework display in each city they visited. In this way Victoria and Albert were entertained in Cologne, Antwerp and Frankfurt during their tour of 1845, and in 1871 sixty thousand people assembled in the Crystal Palace grounds to enjoy the display given in honour of the visit of the Grand Duke Vladimir of Russia to London.

There was, of course, an ever increasing number of small public and private displays, but the contracts for the huge and expensive shows to celebrate national events were heaven-sent opportunities to the manufacturers, not only to increase their profits considerably, but also as means of publicizing their wares. Hence they greeted the victories during the Crimean War with more enthusiasm than most, and the Peace Treaty of 1856 was the signal for a glut of displays in every sizable town in Britain, including no less than four in London, as well as in most French cities. For English manufacturers Queen Victoria proved an even greater asset, since the celebrations which accompanied Her Majesty's Silver and Diamond Jubilees required numerous and lavish displays, not only everywhere in Great Britain, but also throughout the Empire.

Apart from the occasional spectacles marking national events, the most important displays in Britain were certainly those put on at the Crystal Palace by the Brock Company. Charles T. Brock inaugurated the series at the new and popular resort at Sydenham, London, in 1865. The idea was such a success that the company gave regular displays every year until 1910, and then again after the war from 1920 until 1936, when fire destroyed the building. Thus a series of nearly two thousand displays was brought to an end. They had come to be known as 'Brock's Benefits', after an explosion nearly ruined Brock's business' in the 1820's and, as we are told on a contemporary poster, a sympathizer gave Brock 'the gratuitous use of his comodious Ground (in the City Road) to display an exhibition of fireworks for his benefit'. In an effort to popularize the newly opened Alexandra Palace a firework competition was held at the resort in 1876, and among the prize-winners were Pains, Wells and Wilder as well as Brock's, all company names still used to the present day.

The displays of the period, as typified by the early 'Benefits' at the Crystal Palace, bore a significant resemblance to the modern displays in respect of type of fireworks and general display program, and underlined the changes and improvements made in the previous years. Gone was the elaborate scenery and decorated buildings, and gone too was the 'obelisk' or similar central model. Instead, the spectacle was provided by a greater effectiveness from the fireworks displayed and an altogether wider range of items. The addition of metals like aluminum and magnesium, the latter in about 1865 and the former in about 1894, gave fireworks an increased brilliance. Rockets of 1/2lb. and 1 lb. calibre soared higher with more powerful propellants, shells increased in diameter to 8 inches, 12 inches and 16 inches. (It was Ruggieri who first wrote of a shell with lifting charge and projectile contained in one unit in 1812). The potassium picrate 'whistlers' made their first appearance at the Palace. Yet it was the set pieces which were the main attractions of the time.

The Crystal Palace, with its shrubberies and fountains presented a splendid backdrop for a firework display, but the sheer size and magnificence of the setting demanded a performance on the same grand scale. Hence Brock's introduction of the huge pictorial set piece. Using thousands of colored lances mounted on frames, 'fire pictures' up to six hundred feet long and ninety feet high were produced, depicting such epic events as the Battle of Trafalgar, the Siege of Gibraltar and the Eruption of Vesuvius. After 1879 portraits in fire of Royal Personages, visiting dignitaries, politicians and later even film personalities were reproduced. A spectacular variation was the transformation set piece in which a design made up of quick-burning colored lances gave way to a portrait of slow-burning white lances, the white having been obscured by the brighter colors until that point. By very careful choice of compositions and using lances of different lengths, a popular transformation piece was presented in which a rural scene changed from Spring to Summer, Summer to Autumn and then into Winter.

Various other set pieces like fire-wheels up to a hundred feet in diameter, lattice poles, bouquets and cascades became regular items in displays, and, of course, still remain so. Innovations at the 'Benefits' which have disappeared with the very large set pieces, were the 'living fireworks', so popular at the time. In these, live actors, dressed in asbestos and outlined in lancework, wrestled, boxed, enacted scenes and even walked the tightrope. On one famous occasion an actor named Bill Gregory, but listed in the program as Signor Gregorini, was to slide down a wire from the tower to the terrace. At his first attempt he stuck half way down, and had to remain there suspended for the remainder of the display. As Brock notes in his book, 'his remarks left no doubts as to the country of his origin'. It is interesting to find that a member of the Gregory family still works for the Brock company.

While Brock's gained remarkable publicity and deserved acclaim for their big displays at Crystal Palace and elsewhere, other companies in England were also extending their business and influence. Wells (established 1837), Pains (1860s) and Wilder's were among firms which shared an expanding market in this country and the Empire. On the Continent of Europe many of the present-day companies were well established by the turn of the century, and were rapidly extending their arts, techniques and business. For instance, the Lacroix company was established in 1848 to rival Ruggieri in France, Hansson of Gothenberg dates back to 1888, the Hammargren company was started at Anneberg in 1879, and so on.

In this century the number of displays has not decreased, and the big national event, such as the Peace Celebrations of 1946, the Coronation of Queen Elizabeth in 1953, various Independence Celebrations around the Commonwealth, the Prince of Wales's Twenty-First Birthday Party in 1969, still calls for a pyrotechnic show. Moreover, displays are becoming increasingly popular attractions at carnivals, regattas and fund-raising events. Traditional English occasions such as the Henley Regatta or Cowes Regatta would be incomplete without their displays, and fireworks are regular items on the programs of County Shows and Town Carnivals throughout the summer. In the U.S.A., the State fairs usually boast expensive displays; the spectacular 'Setting the Rhine on fire' shows are frequent entries in the calendars of Moog and other German manufacturers; the French Mediterranean resorts regularly entertain their visitors with displays during the holiday season; the Tivoli Gardens in Copenhagen put on shows through the summer after the fashion of the old London Pleasure Gardens.

Exciting and spectacular as the big displays undoubtedly are for the spectators, to the manufacturers they are often more useful as means of publicity than great profit-making items. In fact, most firms depend on their sales of small fireworks through retail shops to accrue any profits, and most firework factories are engaged to a far greater extent in making fountains, small rockets and candles, 'volcanoes' and 'bangers' for the general public, rather than large shells and set pieces for exhibition purposes. Strangely, in most countries where fireworks are legally available to the public, manufacturers spend the whole year producing stocks which are sold during only a few weeks before traditional festivals. In Britain well over ninety percent of all fireworks are sold within the six weeks before the Guy Fawkes Day celebration on November 5th. Similarly the peak sales period in the U.S.A., is just before the July 4th Independence Day festivities. In France the main demand comes prior to July 14th, Bastille Day, and in most European countries firework parties on New Year's Eve are an old-established tradition.

Children in particular get a special thrill when lighting their own fireworks, and unfortunately are sometimes negligent in their handling of what are potentially dangerous objects, if the instructions are not adhered to. Accidents inevitably happen and have led to considerable opposition being directed against the sale of fireworks, or even their manufacture, in almost every country where they are made. Different governments have reacted in different ways to these campaigns. Legislation in the separate states of the United States emphasizes the divergent opinion on the subject. State laws are continually changing, but in January 1969 it appeared that nineteen states prohibited all types of fireworks, and these included the New England states and New York. Seven others, including Florida, prohibited all types, with a few exceptions. Eleven states, including California and most of the Central states from Montana to Oklahoma, allowed only 'safe and sane' type, which apparently condones the use of flares, candles, fountains, wheels, sparklers, etc., but prohibits explosive elements. In the remaining states, including Texas, common fireworks are generally accepted. Local opinion tends to differ considerably, and it is often left to town or local officials to administer the laws as they think best. Some are lenient about enforcing the state laws, and there is much 'smuggling' of fireworks into areas where they are prohibited by law, especially for the July 4th celebrations. Sometimes there is a strict ban on fireworks in local areas where there are no state laws prohibiting them.

In Germany, no person under the age of eighteen is allowed to purchase fireworks, and the Spanish government has banned the sale of all fireworks in the shops. In Britain anyone over the age of thirteen can buy fireworks from retail shops, but there is a growing pressure on the government to increase this age restriction. The British manufacturers have a joint committee which attempts to educate the public in the correct and sensible ways of storing and handling fireworks. Also there are strict laws governing the transportation of explosive material, and regulations concerning the amount of fireworks which retailers may store, the storage conditions and the manner in which pieces may be displayed in shops. The controls within which firework factories must operate are very severe, and have been arrived at over a period of many decades.

As we have already seen, there were many small manufacturers in England by the eighteenth century, ready to supply the needs of the Pleasure Gardens as well as a growing clientele elsewhere. Long before that time squibs and crackers had been made in considerable quantities for the general public to celebrate the traditional festivities, such as Guy Fawkes Day, Queen Elizabeth Day or Saint John's Eve. The diarist Samuel Pepys mentions a firework party with his family after the victory over the Dutch in 1666. Many of the manufacturers came from the continent as religious refugees, although the making of fireworks was usually a spare-time occupation, carried out in their own homes, after a day's work in the silk mills or woollen factories.

It was in these conditions of manufacture that the main dangers of explosions lay.

The situation was made worse after 1695 when the making of fireworks was completely banned following the anti-government riots, which flared up during the Novermber 5th celebrations of that year. In fact, the Act made it clear that 'if any Person shall make or cause to be made, or sell, give, or utter, or offer, or expose to sale any Squibs, Rockets, Serpents, or other Fire-works, he shall forfeit Five Pounds'. Unfortunately, the legislation was not, nor could be, adequately enforced and only served to drive manufacturers to work in more and more dangerous conditions in back kitchens, with no opportunity to extend and improve their technically illegal businesses. There was no diminution in the supply of fireworks for Guy Fawkes Night or the Pleasure Garden performances. Gun-powder, compositions and stars were stored around kitchens near open fires and gas lamps. There was no official control or inspection, since theoretically no fireworks were being made. The ludicrousness of the position was high-lighted when the authorities even engaged persons to break the law when fireworks were needed for public displays. The addition of potassium chlorate to the list of ingredients used in firework mixtures added greatly to the potential risks involved. Inevitably many accidents occurred.

In 1839 three people were lucky to escape death when a spark from an open fire fell on a pile of gunpowder lying on the table, and this in turn set off a barrel of powder nearby. It was recorded that an explosion shook the premises of a Mr. Mortram, when some rocket stars, which were drying in front of an open fire, burst into flame and exploded. The 'Weekly Messenger' of September 4th, 1825 vividly describes the 'dreadful explosion' in Brock's own factory in Whitechapel, London, when 'ten houses were seriously damaged, and over sixty had their windows broken from top to bottom', and 'one poor woman......was so dreadfully injured by the broken glass that she lies in the London Hospital without hopes of recovery'. Apparently two boys had been ramming rockets, when the ramrod struck against the funnel. The friction caused a spark, which fired some nearby gunpowder. Other fireworks were set off, and eventually fire reached the gun-powder magazines. This caused the real disaster, and for a considerable distance was heard 'a sort of rumbling noise as if of some distant thunder, and the next moment a tremendous and deafening explosion followed, and the air was illuminated with lights of various descriptions, and accompanied by continued reports'.

Firework accidents occurred with frightening regularity, and the damage to property was great. Moreover, the lists of persons badly injured and killed increased continually, and these included many unsuspecting 'third parties' who were unconnected with firework manufacture. At last the authorities decided to act. To try and implement the old legislation would have been very difficult and also undesirable, since many people were employed in the industry and also fireworks were in great demand for public displays. So the 'Gunpowder Act of 1860' was put into force after a hundred and sixty-five years of illegal manufacture. The new law sensibly laid down regulations concerning the making and storing of fireworks, and the preparation of compositions. Justices of the Peace were empowered to grant licenses to those who wished to make or sell fireworks, and inspectors were appointed to make regular visits to ensure that the new laws were being kept. The Act was not perfect, but it did encourage many manufacturers to move their businesses out of the back streets, and to set up bigger and safer factories.

C.T. Brock, who started his Crystal Palace displays in 1866, found that the large quantities of fireworks that he now required, necessitated the building of a large, new factory at Nunhead. This was built with special regard to safety, and over the next few years Brock carried out a series of experiments concerning this subject. Many basic conclusions were arrived at, including findings on such important issues as 'the liability of fireworks to ignite by concussion or friction', 'whether twenty yards is ample distance to prevent an explosion in one shed communicating to other sheds situated at the statutory distance', or 'the liability of fireworks to explode en masse if from any cause they should be accidentally ignited'. The experiments were witnessed by a Royal Commission with great interest, and the results formed the basis for the Explosives Act of 1875, in so far as it relates to fireworks.

The results of the Act were immediately seen, as the number of firework accidents decreased dramatically. Yet one important clause had been omitted from the regulations, and this concerned potassium chlorate and sulphur mixtures. Sulphur by itself is not very reactive, and even in a finely divided state is difficult to ignite. Similarly potassium chlorate is a relatively stable compound at ordinary temperature, and only decomposes when heated almost to its melting point of 368 °C. Yet, when mixed together, sulphur-potassium chlorate is unstable and is liable to ignite spontaneously or to detonate at the slightest frictional provocation. Accidents from these causes had been prevalent before the 1875 Act and continued to spoil the new records thereafter. No less than twenty-eight accidents, and eleven deaths, which were due to these causes, occurred in the twenty-eight years after 1875. Ironically, the last of these accidents was recorded in Brock's factory, when a man emptying crimson stars from a canvas tray into an earthenware jar suddenly found that the slight friction involved had been enough to cause ignition. Soon the whole building was alight, and the man suffered severe burning from which he subsequently died. The authorities at last saw the necessity for a complete ban on all sulphur-chlorate mixtures, and legislation to this effect was introduced by an Order of Council in April 1894.

The Acts of 1860, 1875 and 1894 were so successful in reducing accidents that they still form the basis of the laws relating to the manufacture, storing and selling of fireworks in Britain. With regard to the sulphur-chlorate problem, recently British companies wishing to import some types of shells from certain Japanese manufacturers, have been prevented from doing so by the Home Office Explosives Inspectors because chlorate stars have sometimes been found in these shells. The governments of some other European countries, including Germany, France and the Netherlands, have not been so strict on this matter.

All countries which manufacture fireworks have, of course, made their own laws concerning firework making and handling, and in general most have regulations similar to those in Britain as far as the more vital issues are concerned. In spite of the obvious potential danger in working with explosives, the firework industry is now a remarkably safe one, in most countries. Accidents do occur from time to time, and when they do they tend to be rather spectacular ones, especially when in the form of explosions, sometimes with loss of life. In the East, until recent times explosions were a frequent occurence and often tragic. The Madras accident of 1936 accounted for the lives of thirty-nine people, and the explosion in Macao of the same year claimed twenty-three more. Italy, Spain and Portugal are now tightening up on their precautionary measures after a bad history of firework accidents. In France and Germany the record has been good with little loss of life. The United States, in spite of all its precautions where the public is concerned, has a factory explosion rate which is much higher than average, although probably not as high as in Latin America where regulations seem to be very liberal or flouted by manufacturers.

New materials and techniques bring new problems. Some manufacturers are now using plastic containers in which to store stars, other firework units and chemicals. There is some danger that the friction involved, when the lid of the container is removed, could generate enough static electricity in the plastic to ignite the contents. There appears to be no direct evidence as yet that such an explosion has occurred, but the liability is real enough to worry some experts.

Firework manufacture is an art, but for the manufacturing companies it must also be a business. As such it faces the multitude of problems which confront all other businesses in the modern, competitive commercial world. Above all the work must show a profit, and with rising costs of raw materials and transport, increasing wage bills, higher insurance premiums, and the escalating expenses of various over-heads, the firework business in many countries is facing difficult times. In Britain, although the receipts for shop fireworks in 1968 were a record five million pounds, prices still had to be raised before the 1969 season in order that a profit could be maintained. In many countries small firms have, over recent years, been finding it increasingly difficult to fight against mounting economic pressures. Take-overs and mergers have often been answers to problems in an age when the big combine can find the capital necessary for mechanization and organization so as to hold prices down to a competitive level. In Britain the trend is well illustrated by the merging of the Pains and Wessex firms under the big British Match Corporation; by the take over of Wilder's Fireworks by Brock's and of Wells by Schermuly. In the United States the number of small firms has decreased dramatically in recent years.

Brock's Fireworks, founded in about 1700 by John Brock, is the oldest firework company in Britain, but it is no longer owned entirely by the Brock family. Fireworks are still manufactured under the brand name of 'Crystal Palace Fireworks' to remind the public of those famous displays at the resort between 1865 and 1936, which earned the firm a reputation all over the world. The Brock family concern certainly contributed much towards the development of firework manufacture in Britain during the nineteenth and early twentieth centuries, and was much concerned with the establishment of what came to be recognized as the typical British display, with its large set pieces and unhurried formality. Charles T. Brock, as we have seen, also played a considerable part in the formulation of the important Explosives Act of 1875. Brock's have displayed their fireworks all over the globe, the list of occasions being long and varied. Displays for the Prince of Wales's Tour of India in 1875, the Olympic Games celebrations of 1912 and the Royal Tour of South Africa in 1947 are typical entries in the firm's calendar. The company has been situated at Hemel Hempstead in Hertfordshire, and has other works at Swaffham in Norfolk and at Sanquhar in Scotland. Besides display and shop fireworks, the firm also makes signals for the Government, furnace igniters, insecticidal smokes, gas cartridges and bird scarers.

Pains-Wessex Limited was formed in 1965 as the result of a merger between James Pain and Sons Limited of Mitcham in London, and WAECO Limited of Salisbury, Wiltshire, as a subsidiary of the British Match Corporation. James Pain founded his firm in the 1860's, assisted by the advice and experience of his uncle, a member of a well-known family of pyrotechnists called Mortram. After manufacturing on a limited scale, (and suffering several accidents), at Walworth in London, 'Pains Fireworks' moved in 1877, after the Explosives Act, to a new factory at Mitcham. At about this time Pains also built a factory in New York and competed fairly successfully in the American market, producing many part-scenic, part-firework displays, similar to those with which the firm was thrilling the patrons of the Surrey Gardens. The huge frames, which were used to depict the Eruption of Vesuvius or the Battle of Trafalgar before the turn of the century, were to be found in the factory sheds until very recently. Like Brock's, Pains have travelled the world giving displays, especially for Royal visits to various Commonwealth countries and for Independence celebrations of former colonies. Some of the long-standing displays in England include those at the Cowes Regatta and at Alexandra Palace, while smaller shows at various seaside resorts have been regular dates for decades. Success at the Firework Competitions in San Sebastian and Monaco in 1969, 1970 should add to the firm's long-established reputation.

James Pain's father was a William Pain, born in 1806 and described as a Gunpowder maker from Bow. In Pain's recorded history there is a reference to "Mr. Pain who makest the shining gunpowder, liveth at Temple Hill upon Bow River, where he maketh powder for His Majesty's Service. He maketh some also of several prices, and it will be sold by whole barrels and by retail by Mr. Pluett living in York Street, Covent garden, at the Peacock, where he will be found both in the morning and in the afternoon, and at Exchange time upon the French Walk." It seems to be more than a coincidence that the

Pains mentioned here are all related, thus making the Company one of the oldest existing manufacturers.

The company is no longer directed by the Pain family, since after James Pain died in 1926 his nephews Philip and Arthur Milholland took over, to steer the business through the difficult financial period of the thirties and then during the War, when the factory turned to making Very cartridges, wing-tip flares and other items for the war effort. Peace time brought a return to normal firework production and increased prosperity for the firm. The amalgamation with WAECO meant the closure of the old Mitcham plant and the removal of the works to a new extensive and modern factory at Salisbury under the direction of the British Match Corporation. WAECO (Wessex Aircraft Engineering Company Limited) had been founded in 1933 to make wind direction smoke generators for use in civil light aircraft. As a result of the trade depression the firm took up experimental pyrotechnic work with the encouragement of the nearby Porton Government Experimental Establishment. The company prospered, especially under the direction of Mr. E.H. Wheelwright, after 1937. During the War smokes and other items were made for the Government, and later pesticidal smokes (known as Fumite) and fuses for stage and film effects were added to the list of products. In 1947 Wheelwright launched into fireworks under the Wessex brand name, and with new buildings and modern equipment made large strides. The range of products was completed in 1952 when the Wessex marine signal made its appearance.

A third long-established English firework firm still in existence and, in fact, still run by the founders family is Joseph Wells and Sons Limited. The original company was established in 1837 in the City of London. Prior to this Joseph Wells carried on the business of Public Decorators and Limelight Contractors. This consisted of decorating the streets, buildings and gardens with flags, festoons of bunting, ornamental shields and designs, and also little candle bucket lamps and Japanese lanterns. Such service was in great demand for the celebration of weddings and births, and at religious festivals, or for the passage of Royalty or other dignitaries through towns. Before the introduction of the present electric lamps, gas limelights were in constant demand in theatres and concert halls and also for outside floodlighting and illuminations. Early in the nineteenth century fireworks became allied to the decoration business, as they were being called for to give realistic representations in mystery plays, and were employed in pageants and processions held to celebrate great occasions. The invention of colored fireworks greatly widened the possibilities of the pyrotechnic art and thus increased its popularity.

The Wells, business became a private limited company in 1919, then operating two factories, one in South-East London and the other at Colchester in Essex. The latter ceased to exist in 1938, when a modern and spacious factory was built at Dartford in Kent. The London works continued until 1947. Like many other firework firms, Wells made signals during the war, and still make smokes and flares. Members of the fourth and fifth generations of the Wells family are now concerned in the business, but the company has recently been taken over by Schermuly, the London signals manufacturers.

Besides Brock's, Pains-Wessex and Wells there are several other firework companies which were established rather later in Britain, the best known being Standard Fireworks and Lion Fireworks, both based at Huddersfield in Yorkshire.

The firework industry in the U.S.A has been on the decline since the war, and many companies, including some large ones like the famous Unexcelled Chemical Company and United Firework Manufacturing Company of Dayton, Ohio, have ceased production in recent years. The number of companies had increased rapidly from the middle of the nineteenth century. They were founded mainly by English and Italian immigrants, and produced fireworks principally for the 4th of July celebrations. Brock's and Pains competed for the market at that time with some success. Today the number of firms stands at about fifty, but most of them are small, and a large percentage of the business is in the hands of a few principal producers like the Keystone Firework Manufacturing Company of Dunbar, Pennsylvania, the Tri-State Company of Loveland in Ohio and the New Jersey Firework Company of Elkton, Maryland.

Two main reasons have combined to cause a decline in the industry. Restrictive legislation in many states, including the banning of mail order sales, has severely affected many firms. Also the competition from Japan, Hong Kong and Macao has secured a considerable portion of the U.S. market. Low labor costs in these Eastern countries result in relatively low prices, and some of the Japanese effects, especially from shells, provide severe competition to U.S manufacturers. Yet, in spite of set-backs, about sixteen million dollars are spent annually in the U.S on fireworks, including thirteen million on products made there. About sixty percent of the sales are made between

June 28th and July 4th, and most of the rest at Christmas and the New Year.

Many U.S companies are family affairs. Keystone has been owned by the Lizza family since 1912, and Tri-State is controlled by the Rozzis, another family of Italian origin. All companies complain of diminishing profits and blame state regulations as the main culprit. The U.S is also noted for the large number of amateur pyrotechnists who have their own magazine, are supplied with materials by the chemical companies and often operate illegally, and invariably dangerously.

The curious contradictory situation in the U.S.A., where strict state regulations in some states exist alongside seemingly complete pyrotechnic freedom in others, is far removed from the laws pertaining in Russia. There, since the Revolution, the making of fireworks has been carried on entirely in state co-operatives. Many public displays to celebrate state holidays are put on in parks, and certain non-dangerous fireworks are available in shops, especially just before the New Year festivities. Certainly there is long tradition of firework making in Russia, since displays were recorded as early as the seventeenth century. In 1742, for instance, the coronation of Empress Elizabeth Petrovna was celebrated by a big display in Moscow. Tzar Alexander II put on some famous displays in Moscow and St. Petersburg at the time of his coronation, when music and singing from massed choirs accompanied the pyrotechnics.

The Italians have always been famous for their firework displays and must take a great deal of credit for pioneering the art in Europe.

It is said that there are almost 200 firework firms in Italy, though it appears that most of them are small family concerns. In common with many other European firms, those with the larger plants also manufacture signals. Representative firms would be Coccia Serafino of Rome, Camocini of Como, Panzera of Moncalieri. Other well known firework makers are Francesco d'Addario, Mastrodonati Gino, Parente and Fratelli Francano D'Arcangelo.

The first firework manufacturing in Japan was probably about 1620, although displays were performed there before that date. It is said that the word 'Hanabi', the Japanese for firework, was used in 1585 for the first time. 'Hana' means 'Flowers' and 'Bi' is a softened sound for 'Hi', which means 'Fire', and 'Flowers of Fire' is the Japanese name which corresponds to the word 'Fireworks'. It is recorded that in 1613 a messenger from James I, King of England, put on a firework display for Ieyasu Tokugawa, the founder of the Tokugawa government, although it is uncertain whether the fireworks were made in England or China. It is probable too that fireworks were introduced into Japan by the Dutch or the Portuguese, together with fire-arms, at about this period.

In 1659, Kagiya, the famous fireworker, began his work in Edo (now Tokyo). In 1733, the well-known displays on the River Ryogoku in Edo were originated in commemoration of a Buddhist service for the many victims of a cholera epidemic in 1732. The fireworks were fired from ships, and the displays were held more or less annually until 1963.

In 1810, another famous fireworker, Tamaya, left the Kagiya concern and set up his own factory. However, in 1843, the Tokugawa government ordered him to move from Edo, after an outbreak of fire, for which he was held responsible. Fireworks were also developed in many other parts of Japan, under the patronage of feudal lords.

For a long period Japanese fireworks were made only from black powder, but in about 1880 the introduction of potassium chlorate from Europe allowed colored flame compositions to be held. Another milestone in Japanese firework history came when Gisaku Aoki introduced the double-petalled Chrysanthemum in 1926, and two years later, in the firework display to celebrate the enthronement of the Emperor, Aoki displayed a triple-petalled Chrysanthemum, with the centre of the flower in red, the inner petals in blue, and amber round the edge. After this multi-petalled Chrysanthemums were widely developed, and became the country's most representative firework. Japanese manufacturers are also noted for the very large shells which they produce. A thirty-six inch diameter shell takes many weeks to make, and may contain several layers of stars and smaller shells.

A large percentage of fireworks produced in Japan is exported and the bigger firms often put on displays abroad. Marutamaya, perhaps the largest Japanese company, have given shows in places as far spread as Brussels, Moscow, Chicago and Bangkok. Their main factory is at Fuchu on the Tama River near Tokyo, where it has been since the company originated in 1613, and it is still in the hands of the Ogatsu family who were the founders. Other important Japanese firms include Hosoya, which has been established for over fifty years, and Koa.

A large number of well-established firework businesses operate in Germany, many of them making military pyrotechnics as well as display and retail fireworks. The Hans Moog concern of Wuppertal is internationally known and other important companies include Nico Pyrotechnik, J.G.W. Berckholtz, and Flemming, all of Hamburg, Oscar Lünig of Stuttgart, Sauer of Augsburg, Depyfag and, Paul Zink of Cleebronn and Pyro-Chemie of Eitorf. Ruggieri and Lacroix dominate the French firework industry, A.J. Kat of Leyden and Schuurmans of Leeuwarden lead in the Netherlands, Liebenwein is representative of the Austrian manufacturers, Hammargren and Hansson have long been established at Gothenberg in Sweden, Hans Hamberger is well known is Switzerland, Caballer and Brunchu of of Valencia and Pirotecnia Zaragozana and Fernandes Filhos are among several progressive companies in Spain and Portugal, the Barfod-Hoffman family business near Copenhagen has been connected with the Tivoli Garden displays for decades, while Hands of Milton, Ontario, competes successfully with the United States companies.

Long-standing and famous as these companies and many more certainly are, in terms of the actual fireworks produced, there is very little which is new in the manufacture of fireworks across the world. Shells, rockets, Roman candles and mines are the standard pieces produced in every country. Yet, because the history of the industry has been rather different in different areas, because many manufacturers tend to be somewhat conservative and because spectators expect and demand the type of fireworks they have been used to, certain special items or types of display are often characteristic of a particular country. The typical British display tends to be rather leisurely, with time given to absorb the effects presented, and with a number of set pieces, which not so long ago were on an enormous scale. Displays above the Rhine and at the French resorts typify the Continental approach, with short, but very spectacular, shows of predominantly aerial fireworks, many pieces being fired at the same time. Electrical, and hence more controlled firing, extensively employed on the Continent and in the United States, is only just coming into use in Britain, where hand firing still predominates. There is no mistaking a Japanese display with its salvos of Chrysanthemum shells. The 'Crown Wheel' for long a feature of French and German displays is now being seen in shows around Britain. Rockets are the main feature in Spanish displays.

There are a multitude of small differences between the products of manufacturers in different countries and indeed within the same country, which probably few but the expert would notice. The vast majority of spectators at displays will be thrilled by certain items and will take away a general impression of the presentation as a whole. The expert will see much more. He will know through experience what the manufacturer was trying to achieve with a particular firework, what some of the problems would be and to what extent there had been a success. The experienced worker would be able to compare objectively the quality of colors in stars, the efficiency of rockets, or the spread of a shell burst. He would notice whether rockets and shells had burst before the zenith had been reached or after, and whether candle stars had been blown high enough or landed burning on the floor. Pyrotechny is obviously a potentially dangerous pursuit in which few can 'dabble' as in most other hobbies. Hence the number of enthusiasts and experts must be small. Better communications has led to an easier exchange of ideas and information between them, and international firework festivals and competitions, like those at Cannes, Monaco, San Sebastian and Trieste, also encourage this interchange. Where insularity may have tended to encourage characteristic effects and methods of presentation, improved communications should help to diminish this tendency.

In fireworking, progress can be thought of in terms of creating new effects. By experimentation, a new chemical composition can produce a star color not achieved before; the utilization of a metal not used before can produce a new effect, as titanium proved quite recently; a different combination of designs and colors can create an original set piece, and so on. Throughout the history of fireworks, small bands of enthusiasts have brought about a continual stream of new ideas, some successful, and others too dangerous or too expensive to be reproduced commercially. While fireworks remain so popular as the perfect spectacle for celebration or festival, experimentation and progress in the art will, no doubt, continue.

# Chapter 2

# Firework Materials

The materials used in firework manufacture can be divided into the following categories:

(a) Oxidizing agents
(b) Fuels
(c) Color producing agents
(d) Substances which improve particular effects (color, light)
(e) Substances which produce smoke
(f) Binding agents
(g) Phlegmatizers which reduce the sensitivity of mixtures
(h) Stabilizers, which help to prevent chemical reactions
(i) Substances which accelerate or retard combustion
(j) Aids in production, such as solvents, and lubricants

The following list of materials would be more useful if it could be arranged in accordance with the categories mentioned above, but as there is a tendency for functions to overlap, the list has been arranged alphabetically. The list in no way represents a full list of materials used over the wide pyrotechnic range and represents only the materials commonly used for firework manufacture.

*Aluminum*

Over the last seventy or eighty years aluminum has added tremendously to the brilliance of fireworks, and yet the great variety in production techniques has caused problems in the production of uniform effects. Ellern has covered the various grades of aluminum very well in his two works (5) but there is a little more to be said to supplement the specialized nature of this book. The powders are prepared in hammer mills, in ball mills, or by atomization. The first two techniques produce the so called "flake" powders and hitherto have been the ones most frequently used for fireworks. Powders made by this method are stamped or milled with stearic acid or other lubricants so that they form tiny flat plates of irregular shape and large surface area. Foil is used as a starting point and can end up in particles as small as 2 $\mu$ and finer. This type of aluminum is used also for making paint and is known as aluminum bronze, though paint aluminums can contain as much as 3% or 4% grease. The resulting "bright" or "brilliant" silver powders are used in the slower burning silver effects and are usually 120-200 mesh. Three samples of this powder from English sources were in accordance with the following:

| | | | |
|---|---|---|---|
| Retained on 120 | 10 | 10 | 3 |
| 120-200 | 47 | 43 | 32 |
| 200-325 | 26 | 30 | 30 |
| Pass 325 | 17 | 17 | 35 |
| Grease content | 0.03 | 0.25 | 0.4 |

Flake aluminum in larger mesh sizes is known as "Flitter" and there has been a tendency for manufacturers to sell this to the firework trade under the three categories of "fine", "middle" and "coarse". It has to be admitted that as a general rule, a wide variation in mesh size is permissible with flitter but in some compositions it makes a critical difference. The unfortunate variation from one company to another is well illustrated in the following examination:

| | English Sample A | | | English Sample B | |
|---|---|---|---|---|---|
| | Fine | Middle | Coarse | Middle | Coarse |
| 5- 10 | — | — | 5 | — | 10 |
| 10- 20 | 2 | — | 54 | 45 | 55 |
| 20- 40 | 32 | 7 | 41 | 30 | 35 |
| 40- 80 | 30 | 78 | — | 20 | — |
| 80-120 | 22 | 17 | — | 5 | — |
| Per 120 | 14 | — | — | — | — |

| | German Sample | | | Swedish Sample | |
|---|---|---|---|---|---|
| | Fine | Middle | Coarse | Fine | Coarse |
| 5- 10 | — | 2 | 22 | — | 100 |
| 10- 20 | — | 45 | 66 | 59 | — |
| 20- 40 | 68 | 53 | 12 | 28 | — |
| 40- 80 | 30 | — | — | 13 | — |
| 80-120 | 2 | — | — | — | — |
| Per 120 | — | — | — | — | — |

Naturally, with such wide variations in mesh size, there will inevitably be considerable fluctuation in aluminum samples because fine and coarse particles may tend to separate in the drums. As a general rule the following mesh sizes would appear to be the best for flitter:

| Coarse | 5- 30 mesh |
|--------|------------|
| Middle | 30- 80 mesh |
| Fine | 80-120 mesh |

According to Ellern, stamped powders are less common in the U.S.A. and have been replaced with those made in ball mills. This is not the case in England, where the firework manufacturers prefer the denser stamped powders that are a little easier to handle and more consistent in quality.

The so called Dark Pyro Aluminum consists of a very fine flake powder that is produced in varying shades of dark grey, and although it has a nominal mesh size of about 200, it contains particles as fine as 2 $\mu$.

It has been said that this powder is made from burned paper-backed foil, or foil that has been sprayed with varnish which is subsequently charred. The writer has not met this practice but the British Metal Powder Co. of Bolton certainly incorporate a small percentage of carbon black in their pyro-powder. Large quantities of firework aluminum of good quality are produced by the Germans who also use the term Bronze and Flitter; they also use the term *schliff* for flakes and metal inks and *pyroschliff* for the finest firework grades.

Atomized aluminum is being used increasingly in fireworks, but up to the present time only limited use has been found for this material. The problem mainly stems from the fact that it is more difficult to light a spherical or spindle-shaped particle than it is to ignite a flake. One English production method seen by the writer consisted of the dispersion of molten metal, which was poured into a stream of compressed air. The fine metal particles were then carried along into a large collecting chamber. Powder produced in this way certainly varies in shape and is not so regular as some of the spherical grades described by Ellern.

After extensive trials with atomized material, the writer has not been able to utilize any such material coarser than 120 mesh in firework compositions of the normal type (i.e. excluding thermites and incendiaries.) Beautiful effects can be obtained with powders such as 120-Dust or 300-Dust and a grade which passes 120 and is retained on 200 mesh, but these are extremely limited and tend only to be useful in gunpowder type mixes. Perhaps time will remedy the situation and make life a little easier for everyone by reducing the dirt and sensitivity which characterizes the flake aluminums.

*Ammonium salts*

With the exception of the perchlorate, ammonium salts do not find any place in modern firework manufacture. The possibilty of the formation of highly unstable and explosive ammonium chlorate by an ion exchange in the presence of water precludes their use. In the past, white smokes have been made with potassium chlorate and ammonium chloride and it has to be admitted that the mixture appears to be reasonably stable, though theoretically this should not be so.

*Ammonium Perchlorate, $NH_4 ClO_4$*

In recent years ammonium perchorate has been used extensively, not only in fireworks for the production of rich blue and red colors, but also in the manufacture of propellents. It can be safely mixed with pure potassium perchlorate but must not be used with chlorates and it is unwise to mix ammonium perchlorate stars with other chlorate stars in the same shell or rocket. This material is usually imported into England as a white crystalline powder of about 120 mesh.

*Anthracene, $C_{14} H_{10}$*

The pure form occurs as fine blue fluorescent crystals which melt at about 213 °C. It is insoluble in water and rather sparingly soluble in most organic solvents. The main source of supply is from the distillation of coal tar, and the impure commercial grades are frequently greenish-yellow owing to the presence of other hydrocarbons.

Anthracene is mainly used in combination with potassium perchlorate to produce black smokes; an oxygen negative mixture is required for this purpose. During World War II the Germans used anthracene (in combination with hexachloroethane, magnesium, and naphthalene) for smoke production. In addition the shortage of shellac and accaroid resin (which come from India and Australia) caused experiments to be undertaken with a view to replacing these substances with anthracene and naphthalene. A German patent (No. 677532) was concerned with this exercise.

*Antimony Sb.*

The powdered metal, known also as Antimony Regulus, comes to the trade as a dark grey powder fine enough to pass 200 mesh and often 300. It is usually prepared by heating the native sulphide (Stibnite) with scrap iron, or with poorer quality ores by burning off sulphur in a reverberatory furnace to produce the oxide which is then reduced with carbon. The metal melts at 630 °C.

Antimony is mainly used to produce white fires in combination with potassium nitrate and sulphur or it is used in combination with aluminum to aid ignition. Antimony is also responsible in part for the well known glitter effect which is basically a combination of gunpowder, antimony (or the sulphide) and aluminum.

*Antimony Trisulphide, $Sb_2S_3$*

The black powder used for firework manufacture is usually the ground native ore, Stibnite, which is mined in Bolivia, China, Hungary, and South Africa. The Chinese powder used in commerce is very fine, usually passing 200 mesh, and has melting point of about 546 °C. As a fuel, its uses are much the same as the metal powder, though it ignites more easily. It has the disadvantage also that it is more dirty to handle than the metal powder.

Synthetically produced material is not usually used in fireworks and it can be difficult to get good glitter effects from it. The red precipitated form of this sulphide can be used for some firework mixtures but this is not common practice.

*Arsenic Disulphide, $As_2S_2$. Realgar*

The native ore, realgar, is sometimes ground to a fine powder and used to make white fires. It has also been used for making smokes. Realgar is also produced by sublimation when arsenopyrite is roasted. It comes from the U.S.A., Canada, France, Sweden, and Britain, and consists of a red powder, soluble in acids and alkalis, with a melting point of 307 °C.

*Arsenic Trisulphide, $As_2S_3$. Orpiment*

This sulphide also occurs in a mineral form. The commercial powder has two forms, one yellow and one red. The yellow form changes to red on heating to 170 °C. It is insoluble in water and hydrochloric acid, but dissolves in alkaline sulfide solutions and nitric acid.

The red form is often used to make white stars which have the advantage of being easy to ignite when moving at very fast speeds. Apart from the occasional use in smokes, orpiment is used in combination with carbon black for making Flower Pots with their characteristic golden spur fire.

Arsenic is safe to handle of course, provided that precautions are taken to keep it out of the nose and mouth. Perhaps the name itself conjures up images of other uses to which some arsenic compounds have been placed, but it should be remembered that soluble barium, for example, is equally toxic.

*Asphalt, Gilsonite*

This is a blackish brown solid imported from Syria, Egypt, and Trinidad. It is possibly formed by chemical changes and the oxidation of high boiling point mineral oils. Asphalt with a melting point of about 100 °C is normally used since the ones with lower melting points are more difficult to keep in powder form. It is insoluble in water and alcohol, but dissolves in coal tar naphtha, turpentine, and petroleum.

Asphalt is seldom used in English firework manufacture but seems to have found considerable use in American formulations. The possible sulphur content would possibly make it hazardous with chlorates and there is the additional disadvantage that it is very dirty to handle and tends to leave sticky deposits on tools.

*Barium Carbonate, $BaCO_3$*

Barium occurs naturally in England as Witherite (Barium Carbonate) or Barytes (Barium Sulphate). The barium carbonate of commerce is made either by precipitation or by the conversion of the natural sulphate into the sulphide and then its interaction with sodium carbonate.

Barium carbonate is no use as a coloring agent, but is often used to reduce acid formation in mixtures or to slow down the speed of some compositions.

*Barium Chlorate $Ba(ClO_3)_2.H_2O$*

The fine white powder of commerce is usually of 99.5% minimum purity and is prepared by the electrolysis of barium chloride. It has a melting point of 414 °C and is invariably imported from France into England.

Barium chlorate is one of the most sensitive chemicals which is used in firework manufacture, but it is difficult to manage without it when deep green colors are required. It is wise to use this substance as little as possible and to use it in combination with other substances which will tend to reduce the sensitivity.

*Barium Nitrate, $Ba(NO_3)_2$*

This is perhaps one of the most useful and stable of the nitrates, but is somewhat limited in use because of its high molecular weight. It is manufactured in England by dissolving the native carbonate in dilute nitric acid or by mixing solutions of sodium nitrate and barium chloride. It is marketed as a fine white powder of about 200 mesh and has a melting point of 575 °C.

Green colors are not very strong when they are made with barium nitrate, though the salt frequently features in compositions made with barium chlorate as the main coloring agent.

More than anything else this substance is used in combination with aluminum powder for the production of silver effects. Silver stars, flares and waterfalls invariably utilize barium nitrate, and the aluminum combined with it is frequently mistaken for magnesium by the uninitiated. Below 1000 °C aluminum burns with a silvery gold effect, and this is characteristic of mixtures of gunpowder and aluminum. Above this temperature silver effects can be obtained with the aid of barium nitrate. Occasionally barium nitrate greens are used in situations where there is a danger of ignition from friction (as in filling green lances with a funnel and wire.

*Barium Oxalate, $BaC_2O_4.H_2O$*

Occurring as a fine white powder made by precipitation, this substance is insoluble in water, but soluble in dilute hydrochloric and nitric acids. The price and the weak flame coloration of barium oxalate make it unsuitable for ordinary firework manufacture, but is is sometimes used in more specialized items in combination with magnesium.

*Barium Peroxide, $BaO_2$*

Although barium peroxide is used in pyrotechnics, it is not suitable for use in fireworks owing to its very reactive nature. It decomposes in water and also at 800 °C, and mixtures containing aluminum are likely to heat up in the presence of water. It is prepared by heating the monoxide in a stream of oxygen and has a melting point of 450 °C.

*Beta Naphthol, $C_{10}H_7OH$*

Manufactured mainly for the dye industry, this substance has found an occasional use as a fuel in colored stars, mainly because of its carbon content. It melts at 122 °C. Unfortunately this substance irritates the mucous membranes and is not pleasant to handle.

*Boric Acid, $H_3BO_3$*

This very weak acid is sometimes used in firework compositions to prevent the decomposition of mixtures containing aluminum. Since the decomposition of aluminum is an alkaline reaction, one or two percent of this acid will help to prevent the acceleration of the decomposition. Heat will sometimes accompany this reaction, but the boric acid acts as a buffer, preventing an alteration in the pH. The wet slurry of barium nitrate and aluminum which is used for making sparklers would be quite likely to react if allowed to stand for some time unless there were some buffering.

In dry mixtures, zinc oxide is sometimes added for the same reasons; in this case the oxide is not a buffer, but being amphoteric it may slowly react with any acidity or alkalinity present.

### Calcium Carbonate, $CaCO_3$

The precipitated form of this compound finds an occasional use as a neutralizer in some mixtures, in Armstrong's Mixture, matches, and snakes made with nitrated pitch.

### Calcium Oxalate, $CaC_2O_4$

A fine white powder, made by precipitation, it is not commonly used in ordinary firework manufacture. It has been employed mainly in signals to give depth of color to mixtures of sodium nitrate and magnesium and in other signalling devices. It is insoluble in water, but dissolves in dilute hydrochloric and nitric acids.

### Calcium Orthophosphate - Tricalcium phosphate, $Ca_3(PO_4)_2$

This well-known "flow agent" is used in firework manufacture. About 1% can be added to many mixtures to enable them to move freely in a funnel and wire apparatus, or automatic machines and presses. The white amorphous powder is made by precipitation. It is insoluble in water, but dissolves in acids and has a melting point of 1670°C.

### Calcium Silicide, $CaSi_2$

This material is a grey/black powder which is insoluble in cold water but is soluble in acids and alkalis. It is decomposed by hot water. It finds use mainly as a fuel for self-heating cans of soup and is often an important component in smoke compositions.

### Carbon Black—Lampblack

These blacks are essentially carbon in a very fine state of division, and are prepared by burning oils in such a way that incomplete combustion takes place, so a large volume of smoke is produced; this smoke consists of unburnt particles of carbon and is collected in a specially constructed system of flues.

The quality of blacks vary, but they usually contain about 98% carbon, have a pH of about 8 and usually pass a 350 mesh screen. It is important to make sure that lampblack does not contain traces of unburnt oil which would be dangerous. Carbon blacks, vegetable blacks and gas blacks are all closely related, being made similarly by

burning oil or gas. As the quality varies so much, the firework maker
has to experiment to find the material which suits his purpose.

Carbon black is used to make Flower Pots, the unusual little golden
fountains with their own special type of gold spark. Golden streamer
stars also employ carbon black for the best effects. It is a pity that the
material is so filthy to handle; otherwise it might find a more extensive
use. When compositions are made wet it is advisable to add a proportion of alcohol to reduce surface tension and make it easier to wet
the mixture.

*Castor Oil*

Little use has been made of castor oil in English manufacture, but it
appears to have found considerable use in the USA in the past. It is
used mainly as a protection for magnesium, but it also acts as a binder
or lubricant in that it reduces the friction of the powder against the
walls of the container into which it is pressed.

*Charcoal*

When wood is heated in the absence of air, the volatile products
distil off and charcoal remains behind. For the manufacture of gunpowder certain woods are preferred, such as willow, alder and dog-
wood, but in the manufacture of fireworks it is fairly usual to use commercial grades of mixed hardwood charcoals. The principal English
manufacturer uses a mixture of hardwoods in order to produce a
fairly uniformly blended product.

Three main grades of charcoal are used by the industry, fine, middle
and coarse, though the mesh limits are not accurately defined as can
be seen from the following analysis:

|  | No. 150 | No. 40 | No. 28 |
|---|---|---|---|
| +20 mesh | — | — | — |
| 20– 60 | — | 40 | 76 |
| 60–100 | — | 28 | 8 |
| 100–160 | — | 9 | 9 |
| 160–200 | 5 | 10 | 4 |
| –200 | 95 | 13 | 3 |

Ordinary grades of charcoal have an ash content of about 5 or 6%
and roughly the same amount of water. Better grades of charcoal may
have a 3% ash content. Moisture does not usually present very serious
problems in fireworks, but it is a point to be watched if stores are to
be kept for long periods of time.

Japanese Hemp Coal is produced by burning hemp waste. One analysis showed that it contained 9.3% $H_2O$ and 9.61% ash, therefore the carbon content would probably be less than 81%. Perhaps the material is favored in Japan because it is cheap and very easily reduced to fine powder.

### Chlorinated Rubber. Alloprene. Parlon

This is a white, odorless, granular powder which is sometimes used as a color intensifier because of the chlorine content. Parlon is an American product and not the same as Parlon P, which is chlorinated polypropylene, and which is not used as a color intensifier.

### Clay

This material is an important part of firework manufacture, for it is used to block up the ends of tubes or to provide a washer through which fire can be forced in order to produce pressure. Almost any kind of clay can be used, but it must be thoroughly dried, and sifted to remove stones. When the dried clay is struck a few blows with a mallet and drift or pressed, it forms a very solid mass which does not crumble easily. White fireclay is the type most commonly used.

### Copper Powder

Use has sometimes been made of the bronze and electrolytic copper powders for the production of blue colors, or as intensifiers for green colors, but this is not very common, for the same effects can be achieved by more efficient means.

### Copper Acetoarsenite $(CuO)_3$ $As_2O_3$ $Cu(C_2H_3O_2)_2$—Paris Green

This substance has been known by a variety of names (Brilliant Green, Imperial Green, Schweinfurtergrün,) and it has frequently been wrongly named in firework literature. Copper arsenite (Scheele's Green) and copper arsenate are different substances and less useful. Paris green is prepared by the interaction of sodium arsenite, copper sulphate and acetic acid. It has an intense green color, is insoluble in water and alcohol, but soluble in acids. Needless to say it is toxic, and sometimes causes nose bleeding and skin rashes.

Apart from the compositions employing ammonium perchlorate, Paris green still provides the best blue colors.

### Basic Copper Carbonate

Occurring in native form as malachite $CuCO_3.Cu(OH)_2$ and azurite $2CuCO_3.Cu(OH)_2$, the compound used for fireworks is usually made by precipitation. It is the best copper compound to use in combination with ammonium perchlorate for the production of blue colors, and quite deep shades are obtained in this way. An adequate blue star can be made with potassium chlorate and copper carbonate by adding PVC and cool burning fuels.

*Copper Oxides*

Black cupric oxide was used many years ago (along with the fused sulphide) in the production of blue colors, but calomel had to be used also as an intensifier. These are no longer used. Both cupric and cuprous oxides are used now for ignition and starter compositions in conjunction with silicon and lead dioxide.

*Copper Oxychloride*

This basic chloride appears to have a variable composition and is possibly $3CuO.CuCl_2.3H_2O$. It is formed when cuprous chloride is exposed to air. It is soluble in acids and ammonium hydroxide but not in water. Blue colors have frequently been made with this chloride and it still finds some use since it is cheaper than some of the other copper compounds.

*Cryolite-Greenland Spar. $Na_3AlF_6$*

The production of yellow color in fireworks is something of a problem owing to the liability of sodium salts to absorb water, or to react with other substances and incidentally absorb water. Cryolite is one of the few sodium compounds which is completely insoluble and unreactive. The natural material appears to be better than the synthetic form in many compositions.

*Dextrin $(C_6H_{10}O_5)_n$*

This is the generic name for the products of the partial hydrolysis of starch. As it is soluble in water and has good adhesive properties, it is used extensively as a binder in fireworks. It is fairly usual to add a few percent of the dry powder to a star composition during the mixing operation and then add water prior to star formation. While dextrin is very useful for these purposes, it also tends to be somewhat hygroscopic and so it is unwise to use more than about 5%, and preferably less. Dextrin is insoluble in alcohol; if alcohol is added to the water to promote faster drying during star manufacture, no more than 30% should be added or the adhesive quality of the dextrin will be inhibited.

*Dyestuffs*

The quality of the materials used for smoke production is important, particularly the particle size of the dyes and their freedom from inorganic salts. The sublimation temperature should be as low as possible. The field is very specialized but there is an adequate technical literature.

## Flour

In the past, flour was extensively used to manufacture paste, but most manufacturers now purchase their adhesives. It was also used in some compositions to retard the burning speed.

## Gallic Acid 3,4,5—Trihydroxybenzoic Acid

The acid crystallizes as almost colorless silky needles of the composition $C_7H_6O_5.H_2O$. It is soluble in water, alcohol, and ether, and decomposes at $220\,°C$. The fine commercial powder is used sometimes for the manufacture of firework whistles and is the most dangerous of all the whistle compositions. Being highly sensitive both to impact and friction when combined with potassium chlorate, it has frequently been known to catch fire or explode, when charging operations are taking place. It can not be used with potassium perchlorate in whistle compositions.

## Glass Powder

This is not much used in fireworks, but occasionally is used in match head and striker compositions.

## Glue

Hide glues are invariably used in match heads and certain types of sparklers where a wire or stick is dipped in a wet slurry. Hide glue is also employed in some kinds of cheap smokes to cool the burning process.

Hide and fish glues have always been traditionally used for sticking various firework parts together, but these are gradually being replaced by the more modern PVA emulsion adhesives. These emulsions have the advantages of toughness and a certain amount of elasticity and do not require any heat for application.

## Graphite

The fine black powder of commerce is not used as firework fuel, though it is frequently added to compositions which are to be pressed in molds to ease their ejection.

## Gum Resins

Accroides Resin

Known also as Red Accroides, Grasstree, Yacca, Red Gum
Accroides appears as a recent resin of vegetable origin, to be found

exuding from trees of the Liliaceae family, genus Xanthorroea, which are distributed in all regions of Australia.

Lumps of resin appear naturally about the dead leaf traces around the boles of the trees. Collection takes place during the spring season. The resin is cut off, roughly hand graded and then sieved and separated to remove the fine foreign matter.

The resin is preponderantly Erythroresinotannol but with small quantities of Styracin, benzoic and cinnamic acids, oxycinnamic acid and oxybenzaldehyde.

The brown powder used in fireworks has a wide particle size usually passing 60 or 80 mesh down to 200 mesh. It has a slight bezoin-like odor, melts at about 120 °C and is soluble in alcohols and alkalis but is insoluble in hydrocarbons. The resin is immensely useful in firework compositions and has more or less replaced shellac, particularly as it is so much cheaper.

*Gum Arabic*

Known also as Acacia Gum, this material is obtained mainly from the various species of Acacia trees. Gum Arabic comes to the English market from Aden and Bombay and is mainly of use because it is more or less soluble in water. It occurs in roundish or ovoid pieces of various sizes and varies in color from colorless or pale yellow to brown. The best qualities take up about one and a half times their own weight of water for solution, but the gum is insoluble in alcohol. Alcohol precipitates the gum from aqueous solution.

The solution tends to be acidic, particularly when it has been allowed to stand and ferment and is best avoided as a firework adhesive, particularly in chlorate mixtures. Quite frequently it is used for making quickmatch.

*Gum Copal*

The East Indian Copals, as their name implies, are obtained from Malaysia generally, and in particular from the Malay Peninsula, Borneo, Java, Celebes, Papua, Sumatra and the Philippines. Pontianac is the hardest material readily available, originating from Borneo, whereas the softer variety is known as Manila, Macassar or Papua. These latter copals are won from Agathis Alba by tapping and subsequent collection some three months later. Finally they are sorted and graded according to purity, color and hardness.

The material has a softening point of 80-88 °C and a melting point of 110-125 °C. The powder used in fireworks is light grey in color and has a characteristic smell with a particle size of approximately 60 to 200 mesh. It is soluble in alcohols, esters and alkalis, but insoluble in water.

*Shellac*

Shellac is the name given to the refined form of lac, and the word, derived from shell-lac, specifically refers to refined lac in thin flakes in the form in which it is most commonly marketed.

Lac is the resinous secretion of the lac insect (Laccifer Lacca), parasitic on certain trees principally in India, Burma and Thailand. Large numbers of the tiny red larvae, about 0.5mm long, come out of each mother cell and settle on tender portions of fresh twigs of certain trees, called lac hosts. They feed on their sap, and exude the resin from glands under the skin. The insect completes two life cycles in a year yielding two lac crops. When the material is collected, it is known as sticklac and usually contains about 60-70% of the lac resin and 6-7% of lac wax which is also secreted by insect.

The first part of the refining process consists of grinding and washing to remove coloring matter, sand, insect debris and pieces of wood. The lac resin is then finally produced by melting sticklac in cotton bags and pressing out the melted resin, or by dissolving out the resin with alcohol, filtering and subsequently recovering the resin and alcohol by distillation. The exact nature of the monomeric lac complex, whose molecular weight lies between 1000 and 1100 is still obscure. The present state of knowledge indicates the constitution to be as follows:

1. Dyes.   Two distinct coloring matters are present to the extent of 1/2% (a) Laccaic acid (b) Erythrolaccin
2. Wax.   Lac wax exists as about 4-1/2% by weight of whole lac and consists of even-number primary alcohols from $C_{26}$ to $C_{34}$ and esters from $C_{30}$ to $C_{36}$.
3. Resin.   90% of the total material is resin and consists in the main of hydroxy fatty acids together with their lactones and lactides. Treatment with ether produces two fractions:
   (a) Soluble, known as Soft Lac Resin having an approximate molecular weight of 520, and whose principle component is aleuritic acid, 9, 10, 16-trihydroxypalmitic acid of melting point 100-101°C.
   (b) Insoluble, known as Hard Lac Resin, of approximate molecular weight of 1900 and consisting of about 70% of the lac resin. Its components are derivatives of aleuritic acid.

Shellac used for firework manufacture is an orange-brown powder, softening at 50°C and melting at about 75°C. Powder as fine as 200 mesh is sometimes used, but the normal powder of commerce has a wide particle size range and is approximately 30-200 mesh or 60-200 mesh. Material made by the original machine-made/hot solvent process is frequently used. It is insoluble in water, but dissolves readily in alcohols, organic acids and aqueous solutions of alkalis. Shellac, accaroid resin and copal gum are all used as fuels, mainly for the production of color.

*Gum Tragacanth*

This expensive gum comes from several species of Astragalus growing in Lebanon and Syria. The gum is exported chiefly from Baghdad and Basra. It is soluble in water but not in alcohol. Very occasionally it is used as an adhesive in fireworks.

*Wood Resin-Pine Resin-Colophony Resin-Rosin*

Rosin is the residue of the distillation of turpentine oil from the crude oleoresin obtained from the pine trees, principally from Pinus Palustris and Pinus Caribea. The resin consists mainly of acids of the abietic and pimaric types with the general formula $C_{19}H_{29}$ COOH. Colophony is soluble in alcohol, ether and benzene and has a melting point between 100 and 150°C.

The resin is used in blue colors and occasionally in smokes, but it is difficult to powder, for although it is brittle and straightforward, it tends to agglomerate on standing. A little starch is sometimes added as an aid to pulverization.

More useful appear to be the resinates which are products of the reaction between colophony and hydrated oxides or salts of certain metals for example, calcium resinate is made by fusing rosin with slaked lime. Resinates have the advantage of softening at higher temperatures.

*Gunpowder*

Although blackpowder has long been superceded as an explosive, it still remains as the basis of the firework industry. It is comparatively cheap and is characterized by easy ignition, versatility and excellent fire transfer properties at a wide range of temperatures and pressures. Gunpowder is quite stable chemically; samples of powder which have been found lying for extremely long periods have been found to be practically unaltered in chemical composition.

The material used for fireworks usually has the composition: potassium nitrate 75%, charcoal 15%, sulphur 10%; but other powders are made in the proportions 65, 20, 15 or 60, 25, 15. Apart from the chemical composition, the number of hours of milling and the size of grains are of importance.

In the manufacture of gunpowder, the charcoal and the sulphur are ball-milled together, which forces the sulphur into the porous structure of the charcoal. After this process the potassium nitrate powder is added and all three are ground together in a mill of the edge-runner type. In this mill two heavy rollers rotate in a shallow pan. The rollers weigh several tons. Two hundred and fifty pounds of powder are ground at once, and a little water is added in order to reduce the dangers of grinding dry powder. The milling process lasts from one to seven hours according to the quality of the powder and, at the end of the specified time, the "millcake" is removed from the mill and pressed in a hydraulic press.

The next operation is to granulate the cake by passing it through a series of rollers which break it down to grains of the required size. The grains are then freed from dust by passing the powder through a rotating cylindrical sieve. At this stage the powder is partly dried and glazed by putting it in a rotating wooden drum which in some cases also contains a little graphite to give a polished surface. The powder is finally dried very carefully. Unfortunately charcoal can absorb up to 6% moisture and so it is permissible for blackpowder to contain as much as 1.2%.

Blackpowder in the United Kingdom is manufactured exclusively by the Imperial Chemical Industries who produce a number of grades of which those most of interest in fireworks are the following:

| Type | Grade | Mesh Size |
|------|-------|-----------|
| Meal A | B | 40/Dust |
| NPXF | B | 26/150 |
| FO/A | B | 26/60 |
| F | C | 16/24 |
| FFF | C | 24/70 |
| TP Cannon | C | 8/20 |
| TS 2 | D | 24/40 |
| Sulphurless Meal | | 150/Dust |
| SF G 12 | | 8/16 |

The mesh sizes quoted are typical and are sometimes altered according to requirements. The powder grade is concerned with quality and refers to the milling time which increases in the order B, C, and D.

The following information concerning American gunpowder grades was kindly supplied by E.I. Du Pont De Nemours & Co. The milling time for the Du Pont types was not available.

| Type | U.S.Sieve Numbers |
|------|-------------------|
| FFA | 4/12 |
| FFFA | 10/16 |
| FFFFA | 12/50 |
| FFFFFA | 20/50 |
| FFFFFFA | 30/50 |
| FFFFFFFA | 40/100 |
| Meal D | +50 |
| Fine Meal | +100 |
| Extra Fine Meal | +140 |

Sulphurless powder consists of potassium nitrate 70.5% and charcoal 29.5% and is manufactured in much the same way. It is used mainly in priming and igniter compositions, particularly where they come into contact with magnesium. The sulphur in the ordinary blackpowder coupled with the moisture in the charcoal tend to accelerate the decomposition of magnesium.

A few firework manufacturers in other countries make their own mealpowder by simply ball-milling the necessary ingredients with a little water. The operation is somewhat hazardous, but produces a low-grade powder quite adequate for ordinary firework mixes.

*Hexachlorobenzene, $C_6 Cl_6$*

Extensive use of this material as a chlorine donor in color mixtures appears to have been made in the U.S.A. in the past. According to Ellern, it has now been superceded. Little use of the material has been made in Europe since PVC has been preferred. Hexachlorobenzene occurs as white needles which melt at 229 °C. It must not be confused with Lindane, gamma - benzene hexachloride $C_6H_6Cl_6$ which consists of white crystals with a slight musty odor, melting at 112.5 °C. This substance is the well known insecticide.

*Hexachloroethane, $C_2Cl_6$*

This is a white crystalline powder with a slight camphor-like odor. It melts and sublimes at 185 °C. The main use of this substance is in the manufacture of smokes, and for this purpose the chemical is mixed with substances such as aluminum, zinc, oxide, calcium silicide etc. Unfortunately it is very volatile at room temperature and therefore the chemical itself and the smoke compostion made with it can only be kept in sealed tins.

## Hexamine, $(CH_2)_6N_4$

Hexamethylenetetramine is formed when six molecules of formaldehyde condense with four molecules of ammonia. The white crystals burn with a steady yellow flame and have been used for small indoor fireworks in combination with magnesium and lithium salts. Metaldehyde has been used for the same purpose but has the disadvantage of giving off poisonous formaldehyde during burning. Hexamine would be quite useful as a fuel, but does not appear to have found much application.

## Iron, Fe

There can be little doubt that iron has been used in fireworks from the earliest times. Apart from the steel dust used for the manufacture of sparklers, the best iron for fireworks is the ordinary cast iron borings which have been broken down to a rough powder which will pass 20 mesh. The long needle like fragments give the best effects.

Unfortunately it is only too well known that iron will not keep in firework mixtures and it therefore has to be treated. The oldest method is still the best one and employs linseed oil. About 16 lbs of iron is mixed with 1 lb of linseed oil and the mass is slowly roasted in an iron pan until most of the oil is burnt off. Another method is to coat the iron in oil and then allow it dry slowly at room temperature, spread out in thin layers on trays. If preferred, the iron can be coated also with paraffin wax. In some cases where it is very necessary to be sure that corrosion cannot take place easily, the iron is pre-treated with a solution of dibasic ammonium phosphate.

## Iron Oxides

In recent times, the two oxides have been increasingly used in pyrotechnics. The black magnetic form, $Fe_3O_4$ is used in thermite and incendiary compositions and the brown form, $Fe_2O_3$, has been used in first fires and ignition compositions where high temperatures are needed.

## Lead Oxides

The red form, $Pb_3O_4$, and the chocolate colored dioxide, $PbO_2$, have been used in first fires and ignitions for military purposes and are potentially useful for fireworks also.

## Linseed Oil

For many years, boiled linseed oil has been used to coat magnesium powder to protect it from corrosion. Magnesium powder and oil are mixed together and allowed to stand in a warm place in shallow trays for about forty eight hours, before the other chemicals are added. In practice, lithographic varnish is preferred to ordinary linseed oil. Lately there has been a tendency to replace these oils with polyesters or to use no coating at all, but there can be no doubt that linseed oil renders good protection and its period of usefulness is not yet over. Stores made with magnesium coated with lithographic varnish are good for several years, which is not the case with uncoated magnesium.

### Lithium Carbonate, $Li_2CO_3$

The red coloration of Lithium has no distinct advantage over strontium in firework manufacture and so the high cost of lithium salts preclude their use. An occasional use has been found for lithium carbonate in certain types of indoor firework, some of which use nitrocellulose as fuel and oxidizer.

### Magnesium, Mg

Because of its high cost and somewhat reactive nature, magnesium is not used very much in commercial items. On the other hand it is indispensable in situations where it is essential to gain high candle power in signal flares. The readiness of magnesium to react with other materials and decompose has been mentioned elsewhere, thus making it essential to choose the other materials carefully and invariably to coat the magnesium prior to mixing. Linseed oil and drying oils have been extensively used in the past for these protective purposes, but in recent years there have been attempts to replace the oil with polyester resins and, while these are very good, they are a little more messy to handle, and pressing has to take place soon after mixing. Experiments have been made with potassium bichromate as a means of protection, but it has not found much favor. The modern practice of pressing magnesium compositions dry in aluminum or paper tubes is a great time saver, but the finished product cannot be guaranteed for a long period of time.

The metal used for pyrotechnics is usually pure magnesium, but small amounts of manganese are sometimes added. Magnesium/aluminum alloys are also employed in varying proportions and are rather more stable.

A rough sieving of the English grades 0 to 5 produced the following result:

|            | 0  | 3      | 4      | 5  |
|------------|----|--------|--------|----|
| 40/80 mesh | 51 | 22.1/2 |        |    |
| 80/120     | 43 | 50     |        |    |
| 120/200    | 6  | 32.1/2 | 47.1/2 | 5  |
| 200/300    |    | 4      | 42.1/2 | 25 |
| +300       |    | 1      | 10     | 70 |

Comminution takes place by a process of atomization when the molten metal is dispersed by gas in an inert atmosphere or by a process of cutting from large pieces of metal. Of course the density varies with the method of cutting. Grades 0 and 3 are turnings, or raspings, made by a process using a rotating cylinder rather like a very large food grater bearing upon a cylindrical billet of metal. The fine raspings are then milled and separated by sieving. Grades 2 and 4 are produced by a process of abrasion. A rotating cylinder 12″ in diameter has continuous strips of wire brushes wrapped around its periphery. Against this rotating cylinder is pressed a slab of magnesium about 2″ thick. The product from this process is called "cut" powder and is again separated by sifting. The magnesium alloy powder used for fireworks is usually a 50/50 alloy with a mesh size somewhere between 20 and 60 or 70. It is sometimes added to small firework fountains to give a silver spark which ignites with a crackling noise.

### Magnesium Carbonate, $MgCO_3$

It appears that this material is useful for making potassium chlorate or perchlorate free-flowing. American smoke compositions employ 3% of the total weight of the chlorate itself. 1% of the light carbonate was not so successful in a lance composition as 1% tricalcium phosphate for promoting free flow in the funnel and wire apparatus, though it is obviously potentially useful as a neutralizer.

### Mercurous Chloride, $Hg_2Cl_2$. Calomel

Calomel is manufactured by the interaction of mercury and mercuric chloride. It is a white, tasteless, insoluble extremely fine powder which was formerly used as a chlorine donor in color mixtures. The fantastically high cost of mercury salts precludes their use in fireworks, but there are good substitutes. Lead chloride has been used for a substitute but it is not used now.

### Naphthalene, $C_{10}H_8$

Naphthalene is produced commercially from coal tar from the fraction boiling between 180° and 250°C. It is white solid crystallizing in plates with a characteristic tar-like odor. It melts at 80°C, is almost insoluble in water, but dissolves readily in benzene. The main use in pyrotechnics is for the production of black smoke. See anthracene.

*Pataffin Oil and Wax*

Paraffin oil finds some use in fireworks and frequently performs more than one function at a time. When it is added to a colored fire mixture, for instance, not only does it help to reduce the influx of moisture and reduce the sensitivity, but it also makes the mixture easier to press. Normally about 1% is used for these purposes.

Paraffin wax is usually used to coat metal powders or to waterproof finished fireworks, which are merely dipped in the molten wax.

*Phosphorus P*

Only the red variety finds any application in the manufacture of fireworks and only then in the production of amorces and match striker surfaces. The production of amorces is a very dangerous and somewhat specialized business and it is quite essential to ensure that dry amorphous phosphorus and dry potassium chlorate never get near each other or they will explode violently when they are subjected to even the slightest friction.

Weingart describes liquid fire rockets which require yellow phosphorus to be placed in the tops of rockets. There is obviously no commercial future in this nor is it particularly recommended!

*Pitch*

Hard pitch, the residue from the distillation of coal tar, is sometimes used in powdered form in the production of cheap smokes used for making drain testers but liquid tar absorbed on sawdust often gives better results. The powder has also found some use in the production of colors; good red and green lances, for example, have been made with pitch as a fuel. On the other hand the possible impurities could be hazardous with chlorates and pitch is not much used. German formulations have sometimes employed Beechwood pitch —Buchenholzpech— and the Japanese have shown a partiality for Pine Root Pitch, a by-product of the distillation of oil of turpentine.

*Polyvinyl chloride, PVC. $(CH_2{=}CHCl)_n$*

The Germans during the late thirties appear to have been the first to use this material extensively as a chlorine donor. It is a white powder or granular substance with a softening point of about 80°C. It is soluble in dichloroethane and cyclohexanone. At about 160°C it begins to decompose with the liberation of HCl.

PVC contains 56% chlorine and is normally used in magnesium

compositions. Normally it is marketed under various trade names such as Vestolith, Corvic, Igelith, Vipla. There is also a double chlorinated PVC marketed as Rhenoflex. It is very much superior to ordinary PVC for the production of barium nitrate/magnesium green stars, but it is also very much more expensive.

Closely related to PVC is polyvinylidene chloride or Saran, a co-polymer - $CH_2=CHCl(15\%) + CH_2=CCl_2(85\%)$, which can be used in similar ways to PVC. It is insoluble in most things but will dissolve in cyclohexanone.

### Potassium Benzoate, $C_6H_5.CO.OK$

Whistling compounds occasionally use this substance in combi-nation with potassium perchlorate. The powder must be at least 120 mesh and needs to be kept very dry, since there is a tendency to absorb water from the atmosphere and the mixture will burn but not whistle.

### Potassium Chlorate, $KClO_3$

One of the most important chemicals used in the firework industry, this material is prepared by the electrolysis of potassium chloride solution, and is frequently imported into England from Poland and Switzerland. It is a fine white powder passing 170 mesh and is normally of 99.5% minimum purity. The melting point is 360 °C. The sensitive nature of potassium chlorate is a problem to the firework manu-facturer, particularly in the presence of sulphur, ammonium salts, and phosphorus, and none of these materials should be used with it. Wherever possible potassium perchlorate should be used in the place of the chlorate, but this is not always possible. There is also a tendency to use it more in commerce because it is cheaper than the perchlorate.

### Potassium Nitrate, $KNO_3$

The modern method of manufacturing potassium nitrate is to mix hot saturated solutions of potassium chloride and sodium nitrate. Sodium chloride crystallizes out first and then the potassium nitrate. Purification then takes place by re-crystallization. It has a melting point of 334 °C and is usually imported into England from Eastern Europe. Two grades are used in fireworks, both 99.8% purity, but vary only in particle size. The fine powder passes about 180 mesh, but some use is made of a coarser crystalline powder of about 60 mesh. Unfortunately the fine powder tends to cake and invariably has to be re-ground prior to use.

### Potassium Perchlorate, $KClO_4$

It is obtained by the electrolysis of cold concentrated sodium chlorate solution, followed by precipitation with potassium chloride. It can be freed from potassium chlorate by digesting it with concentrated hydrochloric acid, with which the perchlorate ahs no reaction. Material of at least 99.5 % purity is used and this is frequently imported to England from France and Switzerland. The fine white powder usually passes 170 mesh. It decomposes at 400 °C. Wherever possible the perchlorate is used in preference to the chlorate, but it still requires careful treatment. Mixtures of sulphur and potassium perchlorate are less sensitive than those with potassium chlorate. In view of this, people frequently add sulphur to stars made with the perchlorate to improve the somewhat difficult ignition, but this is not recommended. Charcoal will usually help the ignition of perchlorate stars without the addition of sulphur.

*Potassium Picrate, $C_6H_2(NO_2)_3OK$*

This uncommon salt is normally only prepared in small quantities for the firework manufacturers. It forms yellow crystals which are slightly soluble in water and which decompose on heating to about 300 °C. As with picric acid and other picrates, it is sensitive to shock, but is reasonably safe to handle if care is taken. The salt is used for making whistles and can be pressed alone or mixed with potassium nitrate, stearine or asphalt. Asphalt is the least suitable as it is rather unpleasant to handle in fine powder. It is worth mentioning that since lead picrate is dangerous, lead ramming blocks should not be used.

*Silicon, Si.*

The dark grey powder of commerce is known as "fuzed" silicon. It usually passes 240 mesh and contains about 94/95 % of silicon. The material is only used in fireworks as an igniter in certain types of fireworks which need a hot slag to initiate the reaction, and for this purpose it is frequently mixed with potassium nitrate and gunpowder.

*Silicon Dioxide, $SiO_2$*

Recent years have seen the appearance of very pure forms of silicon dioxide obtained by flame hydrolysis. Products such as Cab-o-sil or aerosil have an average particle size of about 16 $\mu$ and are used as grinding, sieving and "free-flowing" aids for hygroscopic and other products. They also help to prevent the electrostatic charge of powdered substances and have many other uses such as thickening and thixotropic agents for liquid systems, pastes and emulsions.

Pyrogenic silicas are either hydrophillic or hydrophobic. Hydrophibic aerosil R 972 adsorbs only 0.8 millimol water even at 80 %

humidity. Hydrophillic Aerosil, on the other hand, adsorbs 20 times this quantity.

The material is of extremely low density and flies about the workshop, but there is no evidence that it causes silicosis. Some firework manufacturers are said to use the material for coating iron and for adding to flash powders in the proportion of about 0.1%.

### Sodium Nitrate, $NaNO_3$

Chile saltpeter originating in the caliche deposits comes to Europe for final refinement, but sodium nitrate is also manufactured from synthetic nitric acid by neutralization. The crystals are readily soluble in water and melt at 308 °C. Sodium nitrate is not used normally in fireworks due to its hygroscopic nature, though the degree of water absorption is related to the purity of the salt. In combination with magnesium it is useful for illuminating flares, but the stores have to be sealed so that they do not come into contact with the air.

### Sodium Oxalate, $Na_2C_2O_4$

Prepared commercially by neutralization of oxalic acid, it forms white crystals soluble in water but insoluble in alcohol. It melts at about 250 °C and is normally bought as a fine powder of 120/200 mesh.

There are two main uses for this substance. The first is for the production of yellow colors in combination with potassium perchlorate and suitable fuels. Potassium chlorate is unsuitable as it is likely that some double decomposition will take place with formation of deliquescent sodium chlorate. It is also advisable to use spirit only when making stars with this material since water will tend to accelerate any reactions which may take place. The second use is the production of yellow glitter effects, with gunpowder, aluminum and antimony.

### Sodium Salicylate, $HOC_6H_4COONa$

This substance is made by heating sodium phenate in an autoclave with carbon dioxide. It forms white, lustrous, pearly scales, soluble in water and alcohol. The powder used for making whistles must pass 120 mesh and is used in combination with potassium perchlorate. As with potassium benzoate, the whistles are effective but are liable to deteriorate in storage owing to the gradual absorption of moisture. The absorption of water also seems to be accelerated in the presence of natural resins.

### Starch, $(C_6H_{10}O_5)_n$

Although it is almost insoluble in cold water, starch dissolves in hot water to form a thick solution which can be used as an adhesive. It is frequently used for making quickmatch and some forms of it are used for making stars in Japan. Maize starch is sometimes added to compositions to cut down the burning speed. Flour can also be used for this purpose. Japanese soluble glutinous rice starch is a white or slightly brown powder which dissolves easily in water to form a very adhesive paste. It is manufactured by soaking glutinous rice in cold water, steaming and subsequently pounding the material into a rice-cake. The rice-cake is then roasted and finally pulverized.

*Stearine—Stearic Acid*

The material is probably a mixture of stearic and palmitic acids obtained by hydrolysis from their glyceryl esters. The powder used for firework manufacture usually passes 80 mesh. The main use for this material is for adding it to some compositions which are somewhat sensitive to friction. It can also be used in those fireworks where it is desirable to have a long flame.

*Strontium Carbonate, $SrCO_3$*

Strontium occurs naturally in England and elsewhere as strontianite, $SrCO_3$ or celestine, $SrSO_4$. The natural carbonate is frequently used in firework manufacture and is usually a pale pink, fine powder of about 120/200 mesh. Purer forms are made by boiling celestine and ammonium carbonate or fusing celestine and sodium-carbonate. Strontium carbonate is used more than any other salt, for making red flames and stars, but it also finds an occasional use as a retardant in some gunpowder mixtures. It is insoluble.

*Strontium Oxalate, $SrC_2O_4.H_2O$*

This material is manufactured by precipitation. It is a fine, white insoluble powder, losing its water at about 150°C. It is often used to obtain red color as with the carbonate, but is considerably more expensive. The water content is also something of a drawback. As with the carbonate also, it finds an occasional use in the manufacture of glitter stars with gunpowder and aluminum.

*Strontium Nitrate, $Sr(NO_3)_2.4H_2O$*

Needless to say, only the anhydrous salt is used for firework manufacture. It melts at 570°C and decomposes to the oxide at 1100°C. In many ways this is the best of the strontium salts for it gives a very rich flame coloration, but unfortunately the technical grades will very quickly become damp when mixed with potassium chlorate. The inconvenience of having to dry the salt, and use the fireworks made with it fairly quickly, somewhat limit its use. It is most commonly used in combination with magnesium and PVC for the manufacture of red stars and flares which are pressed dry. It also is used in the manufacture of bengal illuminations and road flares.

*Sugars*

Lactose, $C_{12}H_{22}O_{11}.H_2O$ is sometimes used as a fuel in firework manufacture. The water can be removed only at about 125°C and lactose melts with decomposition at 200°C. Used in compositions which are required to react at low temperatures, it is of use in the manufacture of some blue colors. It also replaces sucrose, which is more sensitive with chlorates. Perhaps the most extensive use is in the manufacture of smokes using organic dyes.

Sucrose - beet or cane sugar $C_{12}H_{22}O_{11}$ - is used only in the production of colored smokes along with potassium chlorate and an organic dye or mixture of dyes. The finely powdered confectioners' sugar is used for this purpose, and it appears to be fairly safe in combination with the chlorate and an excess of dye. It melts at 160°C. Sorbitol, $C_6H_{14}O_6$, with a melting point of 98°C., and mannitol, $CH_2OH.(CHOH)_4.CH_2OH$ of melting point 166°C, have also found some use in special pyrotechnic products, the former in smokes and the latter as a fuel in some signal stars.

*Sulphur*

One of the most important fuels of the firework industry, it is a pale yellow, fairly dense powder, all of which should pass 120 mesh. Some use is also made of a rather coarse powder in the manufacture of pinwheels. During the distillation process and presumably the melting which takes place in the Frasch process in the USA, the sulphur solidifies to a solid mass of rhombic crystals. This mass, soluble in carbon disulphide, is powdered and forms the powdered sulphur of commerce. It must not be confused with Flowers of Sulphur which is a sublimed form produced during the distillation process. The sublimed form always contains some free acid, maybe as much as 0.25%, and is therefore quite dangerous to use.

Sulphur melts at 113°C and ignites at about 260°C. It is in plentiful supply, is non-poisonous, cheap, and safe to handle.

*Synthetic Resins*

Recent years have seen the introduction of urea-formaldehyde, phenol-formaldehyde (Bakelite) and resorcinol-formaldehyde resins.

They are generally used in the form of syrups which can be mixed with accelerators and oxidizing agents, and the resulting mass sets to a hard solid under various conditions of temperature and pressure.

### Thio-Urea, $(NH_2)_2CS$

Thio-urea is a white crystalline solid melting at about 180°C. It finds some use in the production of white smoke in combination with potassium chlorate etc.

### Titanium, Ti

The very fine powder of this metal has been used in military pyrotechnics for some time, but usually in very fine powders which are very dangerous and sensitive to handle. In more recent times, the use which has been made of the metal in the aircraft industry has meant that coarse turnings have been made available at a price which makes their use worthwhile. The turnings are easily ignited and can be used in fountains in the same way that iron is used, and brilliant silver effects are obtained in this way. The metal has great advantages also over aluminum because it is clean to handle and is completely unreactive with water. At present all the evidence seems to show that the coarse material is safer than aluminum and magnesium.

### Titanium Dioxide, $TiO_2$

This oxide is not much used by the firework industry, but occasionally features in some smoke compositions and is sometimes added to waterproof paints.

### Petroleum Jelly—Vaseline

The jelly has sometimes been incorporated into illuminating compositions where it waterproofs, binds and helps to de-sensitize - but little use is now found for it.

### Zinc, Zn

Certain recipes in the old firework books included zinc, but little use is made of the fine powdered variety except in the manufacture of certain types of smokes. Zinc smokes are very efficient and very good but they are somewhat sensitive to moisture and have been known to react and ignite themselves. Most schoolboys know that a mixture of zinc, ammonium nitrate and ammonium chloride will catch fire when a small amount of water is added to the powder.

### Zinc Oxide, ZnO

The principal use for this material is also in the manufacture of smokes. Many of the smokes used in warfare are basically mixtures of zinc oxide, hexachloroethane and calcium silicide. Zinc oxide has also been used by the Germans for stabilizing mixtures of barium nitrate and aluminum in the manufacture of Very Stars.

# Chapter 3

# General Pyrotechnic Principles

The burning speed of a firework mixture is governed by a variety of factors, the three most important being particle size, temperature, and pressure. In commercial practice it is the particle size which produces the greatest problems, mainly because any material can vary from a coarse powder to a powder which is fine and impalpable. Naturally variations exist between these two extremes and chemical batches can vary from one manufacturer to another. Indeed it is even possible to get variations between different batches from a single manufacturer. In practice, materials are bought to a specified mesh size; potassium nitrate for example might be bought as 150 mesh powder.

Originally a 150 mesh sieve had 150 wires to the inch, but so much depended on the diameter of the wire itself that this inevitably led to standardizion. In more recent times greater accuracy has come about by the measurement of the opening between the wires. 1 $\mu$ in this system represents $\frac{1}{1000}$ mm, as in the following table:

| American Standard Sieves. | British Standard Sieves. | Sieve Opening in Micron. |
|---|---|---|
| 12 | 10 | 1680 |
| 18 | 16 | 1000 |
| 20 | 18 | 840 |
| 30 | 25 | 590 |
| 35 | 30 | 500 |
| 40 | 36 | 420 |
| 60 | 60 | 250 |
| 70 | 72 | 210 |
| 80 | 85 | 177 |
| 100 | 100 | 149 |
| 120 | 120 | 125 |
| 140 | 150 | 105 |

| 170 | 170 | 88 |
|-----|-----|-----|
| 200 | 200 | 74 |
| 230 | 240 | 62 |
| 270 | 300 | 53 |

It will be noted that British and American sieves are much the same and are still used in commerce.

It is well known that a firework composition will usually burn much faster with the increasing fineness of the chemicals used. On some occasions the firework maker will use this principle in order to make fast burning stars or explosives, but he may also do just the opposite and employ slow burning compositions in portfires. The burning speed of a pyrotechnic composition is also affected by the surrounding temperature, and so it will be found that as the initial temperature increases, so will the combustion rate. Large quantities of gases are usually produced in these burning processes and although these gases normally escape into the air, the story is rather different if the burning takes place in confinement. In an enclosed space the combustion rate rises quickly and may become explosive.

Turning to the important matter of sensitivity, the firework maker is primarily concerned with the sensitivity to shock and friction. To some extent he is also concerned about ignition temperature, but the sensitivity to shock and friction is of greater practical importance, mainly because firework mixtures are usually more sensitive than ordnance compositions. In general firework mixtures which are more sensitive to friction are also more sensitive to shock, but this is not necessarily the case.

Shock sensitivity is usually determined with a drop hammer. A small quantity of composition is placed on the anvil of the apparatus and a known weight is allowed to fall on the anvil. The weight is dropped from various measured heights until a minimum distance can be recorded at which the material explodes. Several trials are necessary for any degree of accuracy, but the method provides a useful comparison between various mixtures. Sensitivity ratings can be produced in this way, one method, for instance, uses picric acid as a standard with a figure of 100.

Friction sensitivity is sometimes vaguely determined by striking the mixture a light glancing blow with a polished hardwood mallet, but a more accurate method uses two weighted porcelain plates. A small amount of the mixture is placed between the plates which rub against each other. A comparative set of sensitivity figures can be obtained by weighting the plates and determining when the mixture explodes or ignites. The Germans use a mixture of potassium chlorate and gallic acid as a standard with a sensitivity figure of 1. On this system a chlorate color star would have a figure between 2 and 8, a perchlorate and aluminum mix would be about 6, while gunpowder would be about 36 or more.

It can be seen from these figures that it is the chlorate mixtures which are the most sensitive, and there is no doubt that barium chlorate presents the greatest hazard in that compositions made with it are almost as sensitive as mixtures of potassium chlorate and gallic acid. Part of the problem seems to lie in the fact that potassium chlorate, for example, begins to decompose at a fairly low temperature, about 350 °C., and that this decomposition is also accompanied by the evolution of a small amount of heat. Perchlorates and nitrates are much safer in this respect.

The extreme danger of mixing potassium chlorate and phosphorus is almost too well known to comment upon. Similarly chlorates and sulphur present a serious hazard when placed together and are not normally used. Indeed such mixtures were banned by law in England as long ago as 1894. A part of the problem in this case is due to the tendency for sulphur to be acidic and so to accelerate decomposition. An article on the instability of chlorate-sulphur mixtures (21) postulates a trigger reaction caused by the formation of polythionic acids on the sulphur. It is possible that air oxidation produces sulphurous acid on the surface of the sulphur and that this immediately reacts with the sulphur again to form polythionic acids. Evaporation of aqueous polythionic acid or a rise in temperature causes the partial decomposition of the polythionic acid as follows:-

$$H_2S_nO_6 = H_2SO_4 + SO_2 + (n-2)S$$

Sulphuric acid is likely to react with the chlorate to form chloric acid and is thought to be a cause of instability, but sulphur dioxide also reacts with the moist chlorate to form the highly explosive chloric dioxide.

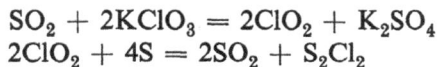

$$SO_2 + 2KClO_3 = 2ClO_2 + K_2SO_4$$
$$2ClO_2 + 4S = 2SO_2 + S_2Cl_2$$

The chain reaction is influenced by friction, impact, sunshine and heat. Experiments seem to suggest that natural sulphur is less reactive than sublimed sulphur and that the formation of sulphur dioxide contributes more to instability than the possible formation of chloric acid.

Some firework makers add sulphur to perchlorate mixtures, and while it has to be admitted that such mixtures are less sensitive than chlorate ones, they are best avoided.

Mixtures of chlorates with magnesium, aluminum, metallic arsenic, the sulphides of arsenic, antimony and phosphorus are all highly sensitive both to shock and friction.

It is clear that nitrates are by far the safest oxidizers, but it is obvious that even they can be dangerous in some situations (e.g. when mixed with fine metal powders).

Very many factors play a part in the ignition of firework compounds but the following general points contribute to an increase of sensitivity:

1. Mixtures with fine particles
2. Intimate mixtures of the chemicals
3. When the components are in stoichiometric proportions, an excess of fuel often decreases sensitivity.
4. Increase in temperature
5. Loose powder is more sensitive than compacted masses.
6. The presence of grit and impurities

Phlegmatizers are frequently added to firework compositions so that friction sensitivity can be reduced; stearine, vaseline and paraffin oil can be particularly useful in this respect. In some cases inert substances can be added and, although they do not take any part in the combustion, they effectively reduce sensitivity. Carbonates can be useful for this purpose and in addition are useful for reducing any tendency towards the formation of acidity. It is advisable for example to add a few percent of barium carbonate to green stars which are made with barium chlorate. Strontium carbonate is particularly useful in red fire compositions as it serves a double purpose of providing color and phlegmatization.

The decomposition of chlorate mixtures require little heat for their initiation and so are unlike either the perchlorate mixtures which do need a little, or the nitrate mixtures which need a large quantity of heat before decomposition can take place. In addition the presence of fine metal powders provides an extra liability to decomposition; mixtures of chlorates and magnesium are particularly hazardous. Potassium perchlorate and aluminum is much safer than the chlorate, as one would expect, but barium or potassium nitrate and aluminum would be the better choice if the required effect could be obtained with these substances, though this is not always possible. Nevertheless barium nitrate and aluminum mixtures present problems and have been known to produce very unpleasant explosions when handled in large quantities. The high ignition temperature of aluminum can make these mixtures difficult to ignite, particularly if atomized aluminum is used, but with the fine flake powders the mixture can be quite explosive. Naturally the addition of sulphur to a mixture of barium nitrate and pyro-aluminum will make the mixture easier to ignite, but it is also considered that this increases the sensitivity and it is best avoided if possible. Wet mixtures of these substances are apt to heat up from time to time and since the decomposition of aluminum is an alkaline reaction, it is advisable to add something to counteract this; one or two percent of boric acid or zinc oxide can be added for this purpose.

Moisture is always a problem to the firework maker and since fireworks have to be stored for periods of time in varying conditions, and invariably in unheated buildings, they are bound to absorb some moisture. When large amounts of water are absorbed physical changes are likely to occur, making the fireworks useless or even dangerous to handle. Chemicals vary in their degree of water absorption, but it is essential to avoid the use of those materials which have a tendency to absorb moisture. Consideration has also to be given to the possibility that in the presence of water electrolytes will exchange their partners to form a partial double decomposition. The chemistry of this reaction is well known, but it would clearly be unwise, for example, to use potassium chlorate and sodium oxalate together since sodium chlorate is so deliquescent. Sodium oxalate can be used much more satisfactorily with potassium perchlorate of course, but all oxalates occasionally present their problems in the presence of water and impure chemicals. This does not seem to be the case however in gunpowder-type mixtures.

It is possible to produce some degree of protection from moisture by the addition of oils and resins etc, but it is essential to be careful in the choice of the most stable materials. Magnesium and aluminum are particularly reactive in the presence of moisture. Fortunately aluminum tends to oxidize only on the surface with the formation of a protective film of aluminum oxide. Magnesium on the other hand does not form such a protective oxide and the corrosion continues. As a rule the magnesium is coated with a protective layer of linseed oil, lithographic varnish, paraffin wax or one of the natural or synthetic resins. Titanium is now becoming increasingly useful and easier to obtain, and has the great advantage of being both unaffected by water and quite safe, provided that its particles are not sub-sieve size. Titanium filings seem safer and more stable than most other metals used though it is very hard and friction sensitivity must be considered.

Since the Second World War the use of magnesium or aluminum with nitrate oxidizers has become increasingly popular, and while this has certain advantages during processing, it has been established that such mixtures are more likely to decompose in the presence of water with the formation of ammonia.

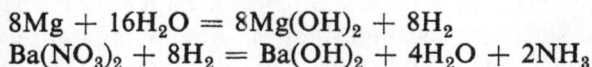

$$8Mg + 16H_2O = 8Mg(OH)_2 + 8H_2$$
$$Ba(NO_3)_2 + 8H_2 = Ba(OH)_2 + 4H_2O + 2NH_3$$

Since moisture is so critical in these mixtures it is clear that barium nitrate is the most useful oxidizer and for much the same reasons aluminum is used much more for firework manufacture than magnesium.

The addition of sulphur to mixtures of barium nitrate and aluminum does not particularly affect stability, but it is unwise to add sulphur to magnesium mixtures as they are liable to react in the presence of moisture. During the Second World War, German igniter compositions for magnesium stars contained sulphur or blackpowder or a combination of these. Problems were encountered with the influx of moisture and an attempt was made to overcome the problem by using a mixture of potassium nitrate, aluminum and gelbmehl (tetranitrocarbazole) (12). According to Schidlovsky (4) powdered alloys of magnesium and aluminum are much more stable chemically for they are coated with a film of aluminum oxide.

From time to time barium peroxide appears in pyrotechnic formulas, but is rather too dangerous for use in fireworks and when mixed with aluminum it is liable to heat up in the presence of water. Ammonium salts need to be used with great caution and must certainly not be mixed with materials which can cause reactions in the presence of water. Potassium chlorate is hazardous with ammonium salts, particularly the nitrate, since the highly explosive ammonium chlorate can be produced by double decomposition. Magnesium is particularly liable to heat up in the presence of ammonium salts, the reaction proceeding at great speed with the evolution of hydrogen. (The writer once had a "close shave" when ammonium perchlorate was accidentally added to a magnesium mix instead of the corresponding potassium salt.) Certain old-fashioned commercial smoke formulas were compounded by mixing ammonium chloride with potassium chlorate and are somewhat unusual because they appear to be relatively stable. In general, ammonium salts are not now used, with the exception of the perchlorate.

Quite the best blue colors can be made with ammonium perchlorate and one of the usual copper salts. A good green can also be made with barium nitrate and ammonium perchlorate, but these mixtures tend to be slower burning and, because of the amount of gas they produce, have been reported to explode occasionally. Perhaps we need to exercise some caution in using this oxidizer, though the fear which some firework manufacturers show is perhaps based on lack of experience and it maybe unjustified. Needless to say it would be foolish to mix stars made with ammonium perchlorate with other stars made with a chlorate and, in firework plants, ammonium perchlorate requires a shed of its own. Basic copper carbonate is the best copper salt to use with ammonium perchlorate for blue colors, but care must be taken not to add either magnesium or aluminum to such a mixture (or indeed to any blue mixture) since the corrosion of aluminum powder is accelerated in the presence of copper or mercury.

The commonest oxidizers used in firework manufacture are the nitrates of potassium, barium, strontium and occasionally sodium; the chlorates of potassium and barium; the perchlorates of potassium and ammonium. It is essential that non-hygroscopic salts should be used and it is not usual to employ sodium or strontium nitrate unless they are very pure. Lead nitrate was occasionally used in former times but it has fallen into disuse. Potassium bichromate is sometimes used in match heads but it has little use in fireworks because it does not yield enough oxygen.

Ellern (5), quoting Schidlovsky, provides a useful table with the relative positions of oxidizers. The following would be of interest to the firework maker:

| Salt | % RH over saturated solution 20 °C |
|------|------------------------------------|
| $K\,ClO_4$ | 99 |
| $Ba(NO_3)_2$ | 99 |
| $K\,ClO_3$ | 97 |
| $NH_4ClO_4$ | 96 |
| $Ba(ClO_3)_2$ | 94 |
| $Pb(NO_3)_2$ | 94 |
| $KNO_3$ | 92.5 |
| $Sr(NO_3)_2$ | 86 |
| $NaNO_3$ | 77 |

A percentage figure lower than 92.5 is hygroscopic thus tending to eliminate sodium nitrate. Strontium nitrate is useful when it is mixed with materials which do not encourage ionization. A good red star for example can be made with strontium nitrate, magnesium and PVC (12). Provided that a good igniter is used, and that moisture is excluded, this star keeps well for a limited period. On the other hand it is unwise to use potassium chlorate and strontium nitrate together.

In general, firework compositions contain one or more oxidizing agents and a fuel or mixture of fuels. Charcoal and sulphur are the most common fuels but extensive use is also made of natural gums and resins such as accroides, shellac and copal. The list is quite a large one, including materials such as lactose, sawdust or woodmeal, starch and dextrine etc. (The reader is referred to chapter 2.) It ought to be mentioned that charcoal and woodmeal can absorb large quantities of water and have to be carefully watched. Dextrine similarly will absorb water, but is used extensively in small percentages as a binder. Aluminum and magnesium are fuels though they also provide special spark or illuminating effects. Iron, antimony, copper and titanium are also quite useful, especially titanium which produces the most beautiful silver sparks and no treatment of the metal is required. Zinc powder is sometimes used in smoke compositions, but it is little used in firework manufacture mainly because it is so reactive with water and various electrolytes.

Colored light is produced by the vaporization of the compounds of certain elements. Red, green and blue colors are produced by the elements strontium, barium and copper and while the halides of these metals would be the best compounds to employ, they are unsuitable in practice. The chlorates, chlorides and perchlorates of these three elements are either hygroscopic or deliquescent (with the exception of barium chlorate) and are quite useless in practice. Strontium monochloride, barium monochloride and cuprous chloride are the three compounds required for the color production and the excess of chlorine has to be present to ensure their formation. Polyvinyl chloride (PVC) chlorinated PVC (Rhenoflex), chlorinated rubber (Alloprene), polyvinylidene chloride (Saran), hexachlorobenzene and mercurous chloride (calomel) have all been used. Calomel is not much used now because of its extraordinarily high price.

For ordinary firework purposes red colors are usually made with potassium perchlorate and strontium carbonate while greens are usually made with barium chlorate. Blue is more of a problem and care has to be taken to keep the flame temperature as low as possible and reduce the oxygen balance. Paris green is often used in blue stars but it is also possible to use copper oxychloride and basic copper carbonate. Yellow effects are normally produced by using potassium perchlorate and sodium oxalate or cryolite. The latter is particularly useful because it is not affected by water. Sodium nitrate would be very useful if it was not so liable to attack from moisture.

The ignition of the majority of the firework compositions is a simple matter because they invariably ignite at temperatures less than 500 °C, and a gunpowder prime or paper impregnated with potassium nitrate may be all that is required. The matter is not quite so straightforward when the ignition temperature is higher, as for example with a mixture of barium nitrate and atomized aluminum. Similarly it can be hard to ignite a highly compacted surface, such as a magnesium star which has been pressed at several tons. In these circumstances it is usual to employ a composition which produces a very hot slag which will ensure the transfer of sufficient heat to cause ignition. Mixtures used for these purposes usually contain potassium nitrate and silicon or lead dioxide, cupric oxide and silicon.

Time fuses and delays for ordinary firework purposes are usually made with compressed gunpowder which burns at the rate of about 2.5. seconds per inch. The so-called Bickford fuses are usually trains of modified gunpowder which are covered with flexible and waterproof covers. Ellern and Schidlovsky have written excellent textbooks in recent years and the reader is referred to these works if he requires to know more of these principles in detail, or to learn of the more complex or exotic materials used in pyrotechnics. (4,5).

The following summary may be useful.

Never mix chlorates with :-    sulphur or sulphides
ammonium salts
phosphorus
pitch or asphalt
picric acid or picrates
fine metal powders

Avoid chlorate and oxalates (moisture problems).

Chlorates and gallic acid are very hazardous. Avoid friction or impact with chlorate mixtures and do not charge them with a mallet.

Avoid using potassium perchlorate with:

> sulphur and sulphides
> phosphorus
> picric acid and picrates

Potassium perchlorate and fine metal powders is hazardous.
The same points concerning friction and impact apply also.
Treat barium nitrate and pyro-aluminum with great respect,
Never use fulminates of silver or mercury.

# Chapter 4

# Mixing and Charging

## Mixing

Firework compositions are usually mixed in a special building; indeed it is common practice to employ several buildings: one for sulphur mixtures, one for chlorate compositions, one for ammonium perchlorate and one for magnesium. Needless to say, even if the operator did not use chlorate and sulphur in the same composition, he would not use the same sieve for the two materials either. Sieves are made usually of copper or brass and are sometimes earthed against static electricity.

Sieving is usually a straightforward matter with gunpowder type mixes, but particular care needs to be taken with chlorates and perchlorates, particularly when fine metal powders are also in the mixture. Explosions or fires do not often occur, but the possibility of an electrostatic spark or friction caused by a fingernail scratching the sieve must not be overlooked.

Fig. 2   Mixing composition (Courtesy Pains-Wessex, Ltd.)

Where hand mixing is employed, large circular sieves about two feet in diameter are frequently used, the mixture being sieved on to a large sheet of paper. Frequent use is also made of a nest of square sieves, one on top of the other. (Fig. 2).

On top of the second and third sieve there is a sheet of cloth to retain the material which has passed through the sieve above. The mixture can then be hand-mixed on the cloth before allowing it to pass through the sieve.

As a general rule sieves of 14 and 25 mesh are used for mixing. Chlorate mixtures are usually passed three times through a 25 mesh sieve, while gunpowder/sulphur mixtures usually pass a 14 and a 25 mesh. Rocket mixtures may be sieved once through a 14 and three times through a 25 mesh screen. It is usual to sieve the potassium nitrate or chlorate (i.e. the oxidizer) first on its own before adding the other materials. This is very necessary as most of the oxidizers tend to agglomerate on standing, unless a flow agent has been incorporated. The word saltpeter (Greek petros - a rock) is perhaps evidence of this. Since a good deal of pressure is needed to force lumps of oxidizer through a sieve, it is very unwise to have fuels mixed with it at the same time, particularly with chlorate mixtures.

There has been an increasing tendency over the last few years to use mechanical mixing methods which operate under remote control. The sieved materials are placed in a special drum made of metal or compressed fibre. After the drum has been securely closed, it is placed in a special apparatus which will cause the drum to turn and mix the con-

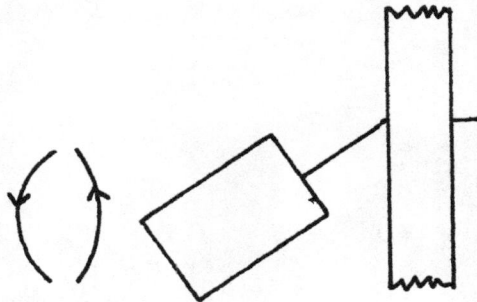

**Fig. 3** Modern mixing drum

tents. Earlier models simply turned end over end, but many people now use an apparatus which holds the drum at 45° to its axis and makes it rotate in a circular fashion. (Fig. 3).

The inside of the drum is frequently fitted with metal baffles but, in some cases, balls of wood, porcelain or rubber are placed in the drum along with the composition.

The whole mixing operation usually takes place in a concrete mixing bay made up of three solid concrete walls, no front, and a light roof extending over the top of the building to exclude the rain. The mixing operation is also usually connected to a gate by electricity, so that the mixing process will stop as soon as the gate is opened. No one can thus be inside the mixing bay during processing.

Open mixers fitted with paddles are also used occasionally, but these are only suitable for mixtures which contain a fairly high proportion of water or other solvent, or for those mixtures which are not particularly sensitive. In all these processes it is important to earth the apparatus against static electricity. Firework makers seem to vary very much in their preferences with regard to mixing methods. In some cases dirty gunpowder and charcoal compositions are mixed mechanically to reduce the amount of dirt and dust and because they are less liable to explosion due to friction. For exactly the opposite reason, they also mix chlorate compositions by hand. However, some would argue that chlorate compositions should be mixed by remote control because they are potentially hazardous. Mixtures of potassium chlorate and gallic acid certainly come into this category and are best mixed in some out of the way place. Mechanical mixing methods can be useful, but they can also be time wasters in situations where a few pounds could be mixed quickly on the spot by hand. There is also a tendency sometimes for coarser materials to settle out, and rise to the surface or sink to the bottom of the mixture. In general it is wise to avoid mixing more than ten to twenty pounds of ordinary composition at one time and, after each batch of mixture is produced, it should be transferred to a suitable expense magazine situated at a safe distance from the mixing shed.

The transportation of mixed powder has problems of its own. Care must be taken to see that the wooden or paper-board containers do not leak, and separate transporters should be used for sulphur mixtures and chlorate mixtures. One accidental factory ignition was caused by sliding a box of blue star composition along a transporter which had been carrying a sulphur mixture. The blue composition ignited but fortunately did not explode. Some companies divide their plants into a sulphur-containing section, often called the "bright" or "black" side, and a chlorate division, often referred to as the "color" side. There is much to commend this practice.

### Charging

Charging methods have slowly changed over the years and fewer and fewer manufacturers use the old-fashioned technique of charging single items one by one, except for special exhibition work. More will be said later about charging methods but the following techniques are available.

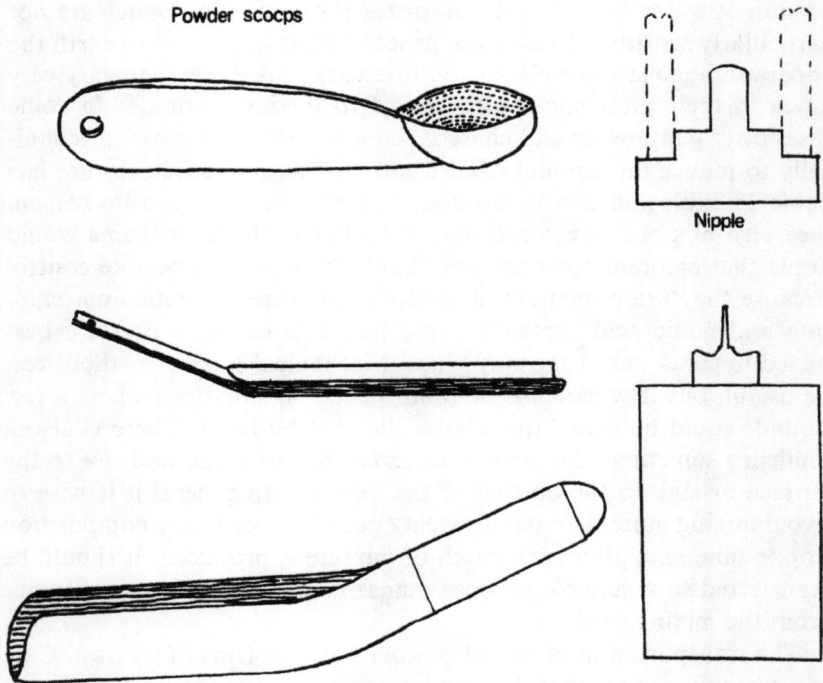

Powder scoops

Nipple

**Fig. 4**  Powder scoops and charging nipples

*Hand Charging*

It is fairly common practice to use a large block of wood as a base for ramming, about 15 inches square and about 30 inches high. In some instances it may be necessary to strengthen the floor of the building also. Mallets usually vary in weight from 1/2 to 10 pounds depending on the nature of the work, and they are usually made of rawhide, box wood, or lignum vitae.

Drifts used for ramming are normally made of box wood, or some of the new synthetic materials, and occasionally of brass. Nipples are normally turned out of a piece of solid brass, aluminum or stainless steel. Powder scoops should be made of copper and are easily made to suit the needs of the charger. The scoops are usually semi-cylindrical and slightly smaller in diameter than the tube which is to be charged. (Fig. 4).

To make it easier to insert the composition, it is often convenient to place a little wooden collar on top of the tube. Where only light charging is required several tubes can be fixed in one frame. (Fig. 5).

The nipples are used for tubes which need some constriction such as fountains and drivers. In those instances where the tube is "pulled in" the nipple merely serves as a base and fills up the end of the tube. Clay also can be used to choke the tube and all that is necessary is to charge a scoopful of clay around the nipple before charging the composition. Mass filling methods are tending to revert to "pulled in" chokes as they are more convenient to handle. Nipples for fountains or drivers choke the tube down from a quarter to a half of the diameter.

*Funnel and Wire*

For charging narrow-bore fireworks, such as pinwheels, lances, and

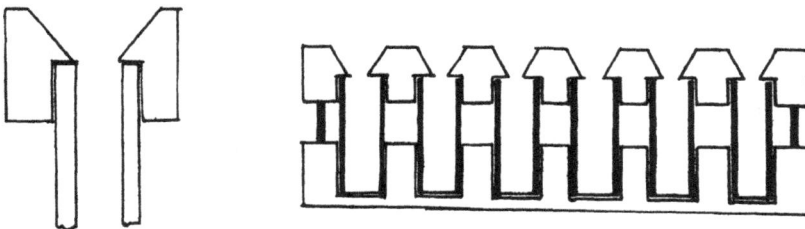

Fig. 5  Single and multiple collars for charging operations

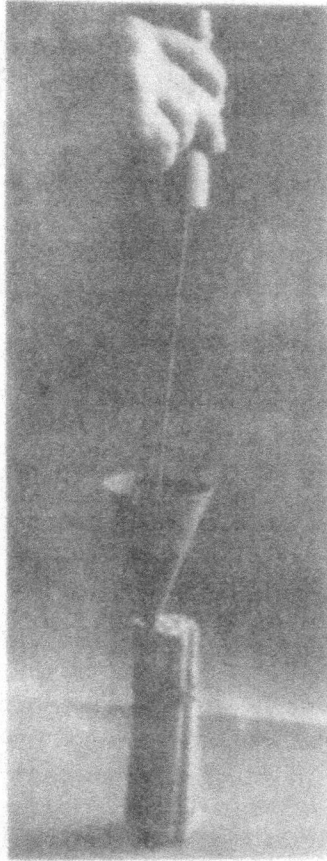

**Fig. 6**   Charging by funnel and wire rod

squibs, a funnel and wire often proves to be the quickest and most efficient method. The narrow end of the funnel is normally turned out of brass, with the remainder of the funnel made of copper which is soldered on to the brass. The narrow brass end usually just fits into the end of the tube to be charged and, after filling, leaves a convenient empty space for the prime. (Fig. 6).

The brass or phosphor bronze rod is as large a diameter as possible, provided that it will allow the powder to flow down the side and into the tube. The wire is usually furnished at the top with a wooden or brass knob or handle. Tubes can be charged quite solidly by this method and it does not take long to acquire some degree of skill. Compositions filled by this method frequently need the addition of a "flow agent", or else the composition will tend to stick in the funnel. 1 % tricalcium phosphate can be added for this purpose.

Many slow burning fireworks such as short squat fountains are filled upside down. The tubes are covered at one end with touch paper and then placed with the touch paper downwards in a large tray. The composition is merely pulled over the open tubes and shaken down by banging the container on the bench. The tubes are then closed up with a paper disc or cork. Needless to say compositions which are filled by this method need to be very slow burning; otherwise a tube filled with a loose mass of fast burning material may explode.

*Filling Machines*

Firework manufacturers tend to develop their own machinery and, having spent a good deal of time and money on this exercise, they are naturally unwilling to pass on the information.

The most common English methods in the past have been those which used automatic ramming machines operated by a cam shaft. Naturally the distance over which the ram can operate depends on the size of the cam, but large banks of these machines can be modified for filling fountains, pin wheels, small rockets and other items. In some instances quite heavy pressures can be exerted by counterbalancing the table on which the tube rests. The machines are also so arranged that they can stop when the tube is full.

Another type of filler uses a shaking table which enables composition to be loosely charged by agitation. The method works well for slow burning compositions, but there can be a tendency for the coarse and fine materials to separate during agitation.

Screw feeding has been tried but there appear to be many drawbacks and a tendency for the machines to get out of alignment and churn up the inside of the tubes!

There can be no doubt that the use of compressed air mechanisms have been a great boon to the firework manufacturer and they appear to be a good deal safer than many other mechanism.

*Presses*

The chief drawback to hand charging is the fact that the operator becomes tired as the day progresses with the result that variations in quality are bound to occur. Presses on the other hand are more precise and controllable, giving good, consistent results. As a general rule it is

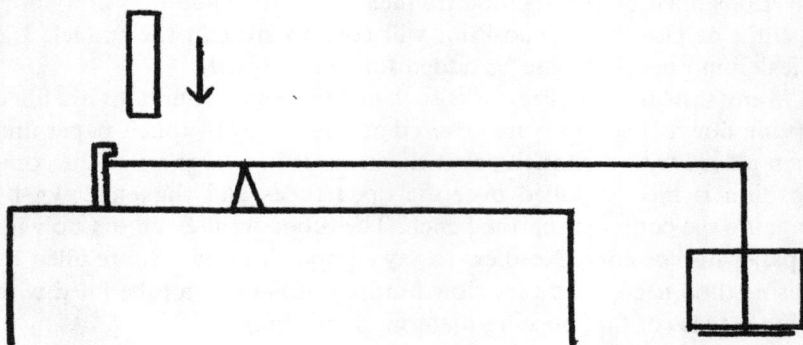

Fig. 7   "Dead load" press mechanism

usual to press sufficient composition to produce a pressed increment
equal in height to the diameter. Pressing loads vary from a few pounds
up to 100 tons or more, indeed the burning characteristics depend to
some extent on the pressing load. In the absence of any other readily
available figures, the following experiments quoted by Dr. David Hart
(11) are of interest:-

Effects of Loading Pressure on Luminous Intensity

| Loading Pressure in pounds per square inch. | Luminous Intensity in candles per square inch. | Burning Rate in inches per minute. |
|---|---|---|
| 6,000 | 78,000 | 9.8 |
| 10,000 | 82,000 | 9.1 |
| 14,000 | 90,000 | 9.0 |
| 18,000 | 93,000 | 9.0 |

Hand presses are useful for work which requires pressing loads up to
half a ton and it is quite useful to fit the press with some kind of coun-
terbalanced table to produce a consistent dead load. (Fig. 7).
Automatic presses are used for heavier loads, though it is usually
necessary to protect the tube with some kind of mold (see Chapter 5).
Pressing without a mold is also possible if the tubes are quite thick-
walled and in these circumstances as many as 120 tubes can be pressed
at one time. The rams are usually fixed to the top of the press or are
arranged to slide along rails. (Fig. 8).

Fig. 8  Press for 321 tubes. (Courtesy Pains-Wessex, Ltd.)

Granulated composition is frequently charged into the tubes by means of a measuring tray with a sliding base. [Fig. 9].

Fig. 9  Measuring tray for composition. (Courtesy Pains-Wessex, Ltd.)

**Fig. 10,11** Part of a modern firework factory. (Courtesy Pains-Wessex, Ltd.)

Loading pressures can frequently present problems mainly because gauges can be inaccurate and vary from one press to another. Naturally it is also necessary to take the diameter of the ram into consideration when calculating tonnage. Conversion from English to Metric systems is as follows:-

1 lb. = 0.454 Kg      1 Kg = 2.205 lb.

1 in. = 2.54 cm.      1 cm. = 1/2.54 in.

$$1 \text{ lb/sq.in.} = \frac{0.454}{(2.54)^2} \qquad 1 \text{ Kg/sq.cm.} = 2.205 \times (2.54)^2 \text{ lb/sq.in.}$$

$$= 0.0704 \; Kg/sq.cm. \qquad = 14.2 \; lb/sq.in.$$

Thus Multiply lb/sq.in. by      Thus Multiply Kg/sq.cm. by 14.2 to
0.0704 to convert to               convert to lb/sq.in.
Kg/sq.cm.

1 Tonne   = 0.984 tons      1 Ton = 1.016 tonnes

A reasonably good fraction for converting the above pressures is 5/71
i.e.   lb/sq.in.   × 5/71 = Kg/sq.cm.
       Kg/sq.cm.   × 71/5 = lb/sq.in.

*Priming*

Fireworks which do not have touch paper ignition are usually primed. This consists of applying gunpowder paste to the mouth of the tube to ensure that ignition takes place. Mealed gunpowder containing about 5% dextrin can be used after it has been damped with water. Some firework makers pour a slurry of prime on to the top of the firework and embed a piece of quickmatch into this. Only a small quantity of prime should be used; otherwise a large amount trapped inside a thick paper capping can be almost explosive. In fact it is often better to use a prime consisting of potassium nitrate, sulphur and charcoal in gunpowder proportions; it is much less fierce.

*Capping and Labelling*

Fireworks for display purposes are generally capped. This means that about three turns of brown wrapping kraft or white sulphite paper are pasted around the end of the tube. The paper overlaps the end by about 1-1/2 inches and is used to tie in the quickmatch pipe where several fireworks are to be fired together. The paper is only pasted at the edge and along the side of the paper which is in contact with the tube. Fancy colored labels are also pasted on the outside edge only; otherwise they dry out on the tubes in a very unsightly manner.

# Chapter 5

# Containers

All fireworks require some kind of container. In the past, tubes have been made out of paper or even wood, but paper is still the most common type of container to be used and is still the cheapest. The last few decades nevertheless have seen many changes in technique, and in some situations where high pressures are exerted, it has been necessary to use metals or resin-bonded tubes instead of paper. Signal flares are frequently pressed into steel tubes, but in those situations where it is necessary for the tube to burn away with the composition, plastic materials or thin aluminum has tended to replace paper. Plastic and aluminum also have tremendous advantages over paper which tends to swell or shrink depending on the water content. Since most of these thinner tubes have to stand up to large pressures during filling, they have to be protected with a mold to prevent the tube splitting. Molds

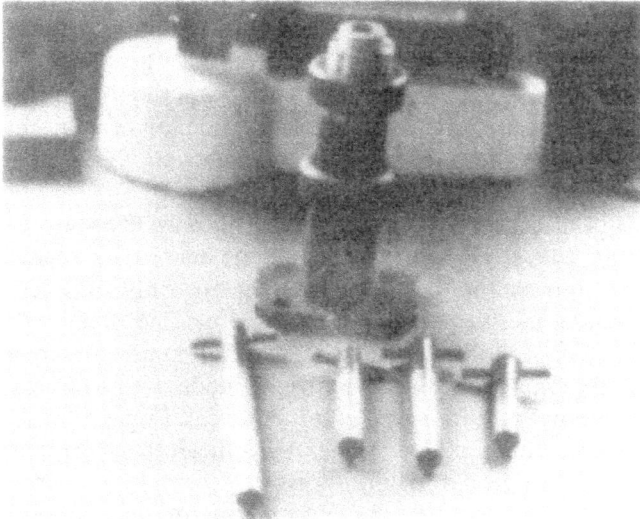

**Fig. 12** Mold and drifts

75

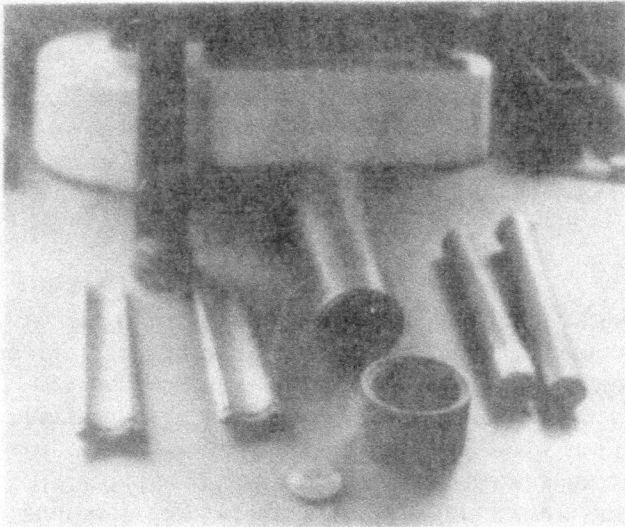

Fig. 13   Sections of the mold

are usually of two types. The first consists of a block of metal into which holes are bored of such a diameter that they will just admit the tubes to be pressed. The inside surface of the molds needs to be highly polished. A plate on the underside of the mold keeps the tubes in place during the pressing operation, and when this is completed the plate can be moved to one side to allow ejection of the pressed tubes.

The alternative method is to use a split mold. This method employs three metal segments which completely surround the tube during pressing. The segments are held in position by a collar of some kind, and there is a taper on the outside of the segments to allow the collar to be removed easily. (Fig. 12 & 13).

Earlier works have expanded at some length on the various ways of rolling paper tubes. No purpose would be served by repeating this information and so the general principles are only outlined here.

Commercial adhesives are many and various and much will depend on the type of paper to be employed. Casein or dextrin based adhesives are frequently employed for strong tubes made with kraft paper, but a starch or methyl cellulose type of adhesive can be used for thinner or cheaper papers. Tubes made with dextrin adhesives unfortunately tend to absorb moisture in damp conditions.

A good cheap paste can be made by mixing one part of wheat flour into a smooth cream with sufficient cold water. Hot (not boiling)

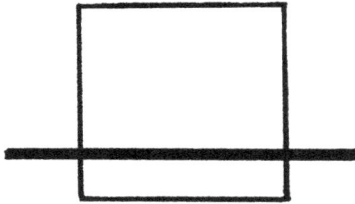

Fig. 14 Rolling narrow bore tubes

water is then added to make the total volume of water up to ten parts. This mixture is then boiled until it thickens. A small percentage of preservative (mercuric chloride, boric acid) should then be added to prevent bacterial action; otherwise there will be mold formation and rotting in damp conditions. The old manufacturers used to add alum as a coagulant, but this is neither necessary nor desirable.

It is not difficult to roll good tubes; it is just a question of the time needed to acquire the necessary skill.

Formers in the smaller bores are usually made of brass or aluminum; larger sizes are invariably brass tubing (Fig. 14). Wood of course is quite useless for wet straight rolling, but can be used for cones or other forms which easily slip off the former. The larger formers are frequently fitted with a wooden handle at one end. (Fig. 15)

Long narrow bore tubes can sometimes be difficult to get off the former. The best way to deal with this is to fix a spike on the edge of

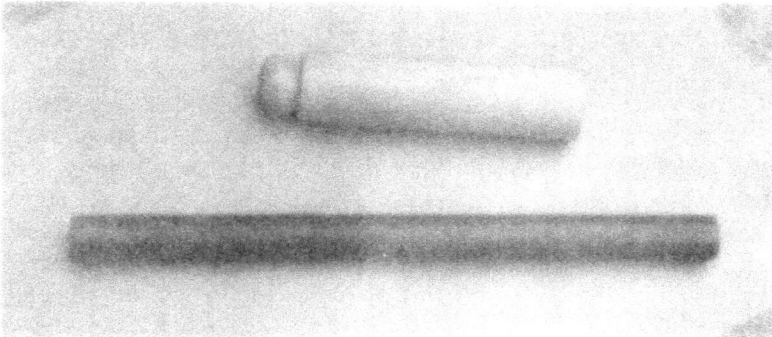

Fig. 15 Formers for wide bore tubes

**Fig. 16**   Using a spike to remove a long tube from the former

the bench and drill a hole through the end of the former. After slotting the hole over the spike, the tube is lightly held as it is pulled off the former. (Fig. 16).

The rolling bench needs to be made of heavy solid timber with a non-absorbent surface which can be easily cleaned. Marble or synthetic resin is very good for this purpose.

The choice of paper is naturally important but not usually as crucial as might be expected. Strong tubes which have to withstand pressure or hammering are usually made with unglazed kraft paper. The weaker tubes for roman candles which only take a light ramming are invariably made of cheap, unglazed, common brown paper (or "sugar" paper, as it is sometimes called).

Paper has a grain and so it is obvious that it must be cut in such a way that the "roll up" is with the grain and not against it. Gently bending the dry paper in the hands quickly shows that it folds more easily one way than the other and needs to be cut accordingly.

In order to obtain tubes with thick walls it is far more convenient to roll several sheets of paper together at once rather than attempt to roll up a single, long sheet of paper. It is often also more convenient to use a large sheet of thin paper on the outside and one or more liners on the inside of a somewhat thicker paper. By using this method it will be seen that it is easier to stick down the outside paper and by using a fairly short "roll up" there is less likelihood of making a bad tube.

The procedure may be best illustrated by an example. Assuming that a tube of 3/4″ inside diameter, 16″ long and having a wall thickness of about 3/16″ is required, two pieces of paper would be needed for each tubes.

e.g.      1 piece      0.014 gauge      16″ by 21″
            1 piece      0.025 gauge      16″ by 11-1/2″

Fig. 17 Paper ready for pasting

If, for example, twelve tubes are to be made, the twelve outside (014) papers are laid on top of each other, a quarter of an inch apart. (Fig. 17).

The quarter inch edges of the paper which are laid bare are then liberally covered with paste and scrubbed with a wire brush to rough up the edge of the paper, to make sure that it will stick down easily on the tube. After this is completed, the batch of papers is turned over and the process is repeated on the reverse side without disturbing the positions of the sheets of paper. As a result of this, each outside paper

Fig. 18 Papers for scrubbing

has its outer edge pasted and scrubbed up, but on opposite sides of each sheet one edge will be on the inside of the tube and the other on the outside of the completed tube. (Fig. 18).

To roll the tube, one of these outer sheets is placed on the bench and covered with paste. Next, an inner (025) sheet in placed about 5″ up the sheet and is also pasted. The former is then placed on the paper; the paper is bent up over the former and checked to be square. After ensuring that the edge of the paper is biting closely under the former, the tube is rolled up, easing the liner a little during the process, if necessary. The edge must be well stuck down and the tube should be rolled on the former three or four times before it is removed. Sometimes a wooden board with a handle is used to roll the tube on the former. (Fig. 19).

When thicker tubes are required, more than one liner can be inserted, though it will be found necessary to space the liners about 1/2″ from each other to give a well balanced wall thickness. The paper should always be uniformly pasted or else there may be the possibility of uneven drying. Drying needs to be a slow gradual process for very wet tubes, a week at room temperature being better than a shorter period at a higher temperature. Wet tubes which have been rapidly dried with ordinary adhesives frequently end up banana-shaped. During the rolling process the tubes become so firmly stuck to the former that they cannot be removed. This is because the rod has become dry, but this can be easily overcome by placing plenty of paste on the former, and the edge of the paper where the former is to lie. Tubes which are used for making lances, pin-wheels or match-piping are not pasted all the way through, the paper being merely pasted on the edge.

Sometimes it is necessary to constrict tubes which are used for rockets, serpents and fountains. The constriction is usually produced

**Fig. 19**  Rolling board

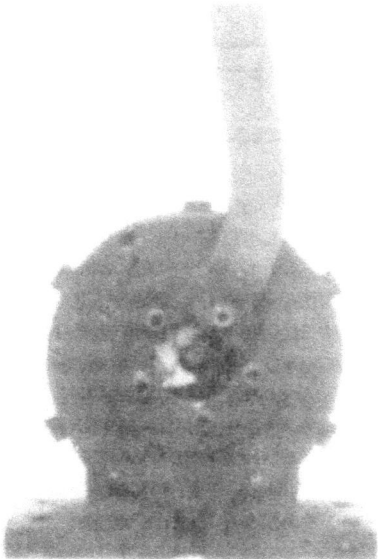

Fig. 20   Choking machine

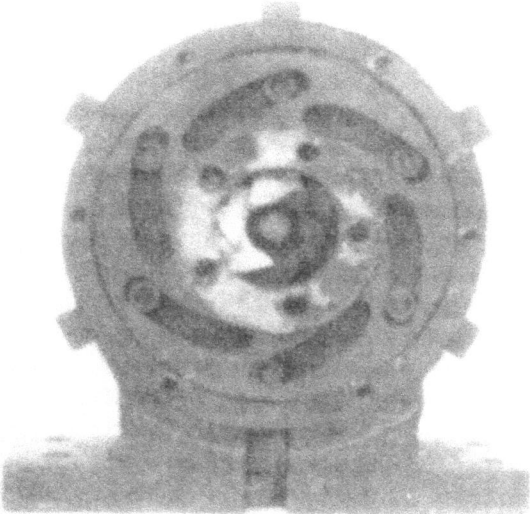

Fig. 21   Interior of a choking machine

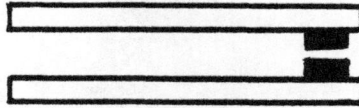

**Fig. 22**   Tube with clay washer

by a machine which contains six plates arranged similarly to those which control the aperture of a camera. Alternatively a piece of piano wire or stranded wire can be wrapped around the tube so that when pressure is gently applied, the tube is constricted. Choking machines can also be used to constrict dry tubes provided that the wall is not too thick, but if the hand method is used, only wet tubes can be choked. Choking machines are also used to close up the ends of small thin-walled tubes. Choking machines are shown in Figs 20 & 21. As chok-ing is a somewhat time-consuming occupation, it is common practice to constrict rockets and fountains with a clay washer. The tube is fitted on to a nipple made of brass, stainless steel or aluminum and a scoopful of clay is hammered or pressed around the nipple before the composition is charged into the tube. At the end of the operation the clay washer is left at the end of the tube after the spindle is with-drawn. (Fig. 22)

Cones and the swell heads of rockets are made by pasting a specially cut paper shape around the appropriate form. (Fig. 23) Cones are

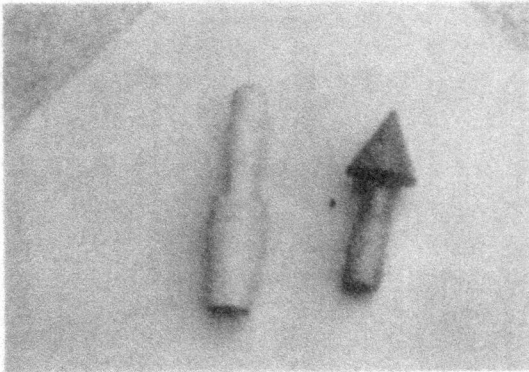

**Fig. 23**   Formers for cones and rocket heads

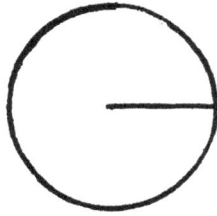

**Fig. 24** Paper circle for cone manufacture

made from a circular piece of paper with a cut across the radius. (Fig. 24)

Wad cutters are used to cut the circular pieces of paper and for cutting the millboard discs which are used for making cylinder shells. Wad cutters are shown in Fig. 25.

Spirally wound tubes are only slowly finding their way into firework manufacture. In those fireworks which do not place a great strain upon the tube they are quite useful, but as a general rule they are not so strong or efficient as the older convolute-wound types. Unless the inside of a spirally-wound roman candle tube has been carefully executed, it is common to find that the pieces of paper are also ejected with the stars.

More accurate details of the tube sizes will be given along with the description of the firework concerned.

**Fig. 25** Wad cutters

# Chapter 6

# Stars

In view of the fact that stars feature in many types of fireworks, their manufacture will be described first. An endless variety of effects may be obtained with the chemicals now at our disposal, though only five types of stars are commonly used; pumped (or cylindrical), cut (or cube-shaped), pill box, pressed and spherical stars.

Before the composition can be made into a star, it has to be moistened with some kind of adhesive so that it will set into a hard mass when it is dry. The amount of water which is added can sometimes be very crucial and care has to be taken not to add too much or too little. It is usual in Europe and the United States to mix a percentage of dextrin to the dry star composition and then add the required amount of water. From 3% to 5% dextrin will normally be adequate and it is unwise to use more than 5% unless it is absolutely necessary, since dextrin is very hygroscopic and stars containing large amounts of it tend to go soft in damp conditions.

Sometimes it is useful to use a mixture of alcohol and water for damping star mixtures; for one thing it reduces surface tension, making it easier to dampen lampblack mixtures for example. It is also easier to dry the stars. On the other hand dextrin does not dissolve in alcohol and so no more than 1/3 of alcohol to 2/3 of water must be used or else the adhesive power of the dextrin will be inhibited. Gum arabic is also used as an adhesive. A 7% solution in water will be found to be adequate, but it can be time-consuming to make up the solution which has to be used the same day. When gum arabic solutions are allowed to stand, they ferment and become acidic and consequently would be very dangerous if used to make coloured stars. Therefore, if it is essential to use gum arabic for some reason, it should only be used on the gunpowder side of the factory.

The other important method of gaining adhesion is to use alcohol only with shellac or accaroid resin. It is common practice to have some of the finely powdered resin in the composition and merely damp the

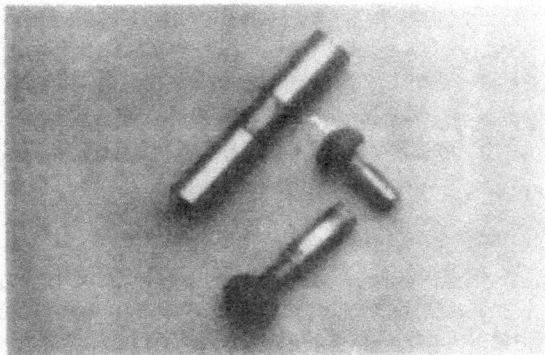

**Fig. 26**  Star pumps

mixture with alcohol, damping a small quantity at one time since it dries out quite quickly. Isopropyl alcohol is the best one to use since it contains no water. It will also be discovered that shellac is the best resin to work with, for it is much less sticky than accaroid resin.

Small pumped stars are usually dried in a gentle heat of about 35°C. for two days. Larger stars and very wet stars will take much longer of course. Drying cupboards are usually heated with hot water pipes and special oil heaters etc. and some manufacturers use vacuum ovens and infra red heaters. In the summer it is often convenient to leave trays of stars out in the open air during the day.

*Pumped Stars*

These are used more than any other type of star in the West, mainly because they can be manufactured easily and quickly.

A single star pump is very useful for making small quantities of stars. It consists of a short cylinder of copper or brass fitted with a brass plunger and wooden knob as shown in the diagram. A stud in the side of the plunger, and a slot cut into the outer sleeve complete the apparatus. When a star is to be made, the plunger is raised to the top of the slot, the pump is pressed into the loose composition and the star is ejected. Some manufacturers use six of these pumps fixed in a row on a wooden handle. The principle is the same except that instead of having slots cut in the sleeve of the pumps, the plunger is spring-loaded. The plungers are normally in the raised position, but the springs allow the plungers to move down the sleeves to eject the stars, afterwards returning to their raised position.

The other common method of making cylindrical stars is to use a star plate. Large quantities can be made by this method, though the stars often tend to be rather soft unless they are made carefully. The star plate consists of a sheet of brass furnished with a number of holes of the same diameter as the star. Underneath this plate there is another plate filled with spiggots, each spiggot fitting into the base of holes in the star plate. The two plates are fixed into the bench and the bottom plate of spiggots has some arrangement attached to it so that the spiggots can be raised to eject the stars. To make the stars, the damp composition is pulled over the surface of the plate and tamped into the molds by another plate filled with spiggots. When the molds are full, the excess composition is removed with a scraper; the base plate of spiggots is raised to eject the stars which are then transferred to a tray.

Roman candle stars have to be very fast and fierce burning; otherwise they frequently blow out of the tube without igniting. For this reason they are usually made with potassium chlorate. Potassium perchlorate can be used of course, but a considerable percentage of charcoal must be added also and the stars always look smaller because there is not so much flame as with the chlorate. The following are good compositions for cylindrical roman candle stars.

| Red | | Green | |
|---|---|---|---|
| Potassium Chlorate | 70 | Barium Chlorate | 53 |
| Strontium Carbonate | 15 | Potassium Chlorate | 28 |
| Accaroid Resin | 10 | Accaroid Resin | 10 |
| Dextrin | 4 | Charcoal 150 mesh | 5 |
| Charcoal 150 mesh | 1 | Dextrin | 4 |
| Blue | | Yellow | |
| Potassium Chlorate | 68 | Potassium Chlorate | 70 |
| Paris Green | 22 | Cryolite | 15 |
| Colophony Resin | 6 | Accaroid Resin | 10 |
| Dextrin | 4 | Dextrin | 4 |
| | | Charcoal 150 mesh | 1 |

| White | |
|---|---|
| Potassium Nitrate | 51 |
| Sulphur | 18 |
| Meal Gunpowder | 15 |
| Antimony Metal Powder | 10 |
| Charcoal 150 mesh | 3 |
| Dextrin | 3 |

Pumped stars usually have diameters of 5/16" or 3/8", 9/16" or 5/8", 1", 1 3/8", 1 1/2". Roman candle stars are slightly less in diameter than the bore of the roman candle tube. Thus for a 5/8" bore candle, the star would be 9/16" diameter and approximately the same length. Gold stars are normally made with charcoal or lampblack, or a mixture of the two. Dr. Shimizu gives details on page 221 of golden streamers made with charcoal only and the following also will be found to be quite useful:-

| | | | |
|---|---|---|---|
| Meal Gunpowder | 54 | Meal Gunpowder | 66 |
| Charcoal 150 mesh | 20 | Lampblack | 23 |
| Lampblack | 13 | Antimony Sulphide | 8 |
| Antimony Sulphide | 6 | Accaroid Resin | 3 |
| Dextrin | 7 | | |

The star containing dextrin is damped with a mixture of two parts water to one part alcohol. The star containing accaroid resin is damped with alcohol only.

The following star is a modification of one quoted by Weingart and is said to be Japanese in origin.

| | |
|---|---|
| Lampblack | 49 |
| Potassium Chlorate | 13 1/2 |
| Potassium Perchlorate | 13 1/2 |
| Potassium Nitrate | 15 |
| Dextrin | 6 |
| Shellac 120 mesh | 3 |

The composition should be made quite wet with alcohol and water like the lampblack stars mentioned above, but they require a long time to dry out.

Very large comets between 1" and 2" in diameter and crossette comets are usually made with a simple mixture such as the following:-

| | |
|---|---|
| Potassium Nitrate | 55 |
| Charcoal Mixed | 25 |
| Sulphur | 15 |
| Dextrin | 5 |

The mixed charcoal is a mixture of all three grades of 150, 40 and 28 mesh according to the effect desired.

A very good comet star is simply a mixture of two parts of gunpowder to one part of fine charcoal, but much depends on how these large comets are to be used in shells.

The well known glitter or twinkling effect often presents problems and some have been puzzled to find that they are unable to get the right effect even though they may have the correct formulation. The problem is very closely tied up with the particle size and method of production of the materials. The antimony sulphide has to be finely ground native stibnite, for the writer has found that the synthetic product is useless for this purpose. The aluminum also seems to be vital and it is usual to use a "bright" powder of about 120 mesh. Some of the German pyro grades work in some formulations, but the best results are obtained with 120 mesh powders made by the stamping process. The glitter effect appears to be produced by the cooling of a blob of dross, produced by gunpowder, antimony and aluminum. Antimony metal can be used in glitter effects but it is not as good as stibnite and some star compositions containing antimony metal have been known to heat up during the damping process. There are two glitter effects, a yellow and a white,

| *Yellow Glitter* | | *White Glitter* | |
|---|---|---|---|
| Meal Gunpowder | 70 | Meal Gunpowder | 66 |
| Sodium Oxalate | 10 | Strontium Oxalate | 8 |
| Antimony Sulphide | 8 | Antimony Sulphide | 14 |
| Aluminum "Bright" | 7 | Aluminum "Bright" | 8 |
| Dextrin | 5 | Dextrin | 4 |

The following is a cheap glitter star for small stars and large comets. It will sometimes function when others will not, depending on the materials.

| | |
|---|---|
| Potassium Nitrate | 32 |
| Meal Gunpowder | 22 |
| Antimony Sulphide | 19 |
| Sulphur | 8 |
| Aluminum "Bright" | 8 |
| Charcoal 150 mesh | 4 |
| Dextrin | 4 |
| Barium Carbonate | 3 |

Older glitter formulations replaced a part of the antimony sulphide with red orpiment. The effect was not particularly improved, and the use of orpiment alone, instead of antimony sulphide, is definitely inferior. These facts and the poisonous nature of orpiment have probably led to the use of antimony sulphide only.

Good yellow glitter effects can be made with meal powder, sodium oxalate, sulphur and aluminum alone. The sulphides of antimony and arsenic are not therefore essential for all glitter effects.

It is difficult to make silver pumped stars which are safe. Very good stars can be made with potassium chlorate and aluminum, such as the following:-

| | |
|---|---|
| Potassium Chlorate | 56 |
| Aluminum "Bright" | 19 |
| Aluminum Dark Pyro | 19 |
| Dextrin | 6 |

but they are sensitive and dangerous. A much safer and more beautiful star can be made with potassium perchlorate and aluminum, but these often refuse to ignite and are best made as pill box stars. The following is an example of a cheap electric streamer without chlorate, but it lacks the brilliance of the chlorate/perchlorate stars.

| | |
|---|---|
| Potassium Nitrate | 45 |
| Aluminum dark Pyro | 30 |
| Aluminum atomised 120 mesh | 10 |
| Sulphur | 10 |
| Meal Gunpowder | 5 |
| Dextrin | about 5% |

The reader is also referred to Dr. Shimizu's discussion of silver star effects on page 224.

*Cut Stars*

Dr. Shimizu on page 251 has described the manufacture of cut stars and little more needs to be added as this corresponds more or less with western practice. Considerably more water has to be added to cut star composition so that it will beat down into the frame in a tacky pliable mass like pastry. If it is too dry, the mass will not cut into cubes and just crumbles. After cutting, the stars are usually agitated in a bowl of mealpowder which coats the exterior.

Some people maintain that the sharp corners improve the chances of ignition of these stars, but this is doubtful. The gunpowder coating undoubtedly improves ignition, but it also increases the danger because of the sulphur and the chlorate being in close contact. Many of the pumped star compositions are suitable for cut stars, but the following are specifically used for the cut variety:-

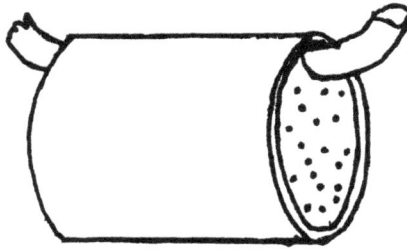

**Fig. 27**  Pill box star

| *Red* | | *Yellow* | |
|---|---|---|---|
| Potassium Perchlorate | 70 | Potassium Perchlorate | 70 |
| Strontium Carbonate | 15 | Sodium Oxalate | 14 |
| Accaroid Resin | 9 | Accaroid Resin | 6 |
| Charcoal 150 mesh | 2 | Shellac 80 mesh | 6 |
| Dextrin | 4 | Dextrin | 4 |

| *Green* | |
|---|---|
| Barium Chlorate | 72 |
| Accaroid Resin | 12 |
| Charcoal 150 mesh | 8 |
| Dextrin | 4 |
| Barium Carbonate | 4 |

*Pill Box Stars*

Although these stars are rather more time-consuming to manufacture, some of the best effects can be obtained with them. Not only do the stars have a longer burning time, usually about 5 seconds, but they always ignite because of the piece of quickmatch which is embedded in the composition.

The stars are made in thin walled paper spools of various sizes, a common size being 9/16″ inside diameter, 1/2″ long and a wall thickness of about 1/32″. The match is inserted into the tube and the damp composition is firmly pressed into the tube with the thumb and first finger. The match extends over the outside edges of the tube by about 1/2″.

The compositions are somewhat slower burning than pumped stars, the following being fairly typical:

| *Blue* | | | *Blue* | |
|---|---|---|---|---|
| Potassium Chlorate | 70 | | Potassium Perchlorate | 39 |
| Paris Green | 20 | | Ammonium Perchlorate | 29 |
| Shellac 60 mesh | 10 | | Copper Carbonate (basic) | 14 |
| add alcohol | | | Accaroid Resin | 14 |
| | | | Dextrin | 4 |
| | | | add water | |

| *Red* | | | *Green* | |
|---|---|---|---|---|
| Potassium Chlorate | 64 | | Barium Chlorate | 48 |
| Strontium Carbonate | 19 | | Potassium Chlorate | 17 |
| Accaroid Resin | 13 | | Accaroid Resin | 17 |
| Dextrin | 4 | | Barium Nitrate | 11 |
| add water | | | Barium Carbonate | 4 |
| | | | Dextrin | 3 |
| | | | add water | |

| *Amber* | | | *Silver* | |
|---|---|---|---|---|
| Potassium Perchlorate | 60 | | Barium Nitrate | 55 |
| Sodium Oxalate | 26 | | Potassium Nitrate | 13 |
| Shellac 60 mesh | 14 | | Aluminum Dark Pyro | 21 |
| add alcohol | | | Dextrin | 6 |
| | | | Sulphur | 4 |
| | | | Boric Acid | 1 |
| | | | add water | |

| *Silver* | | | *Silver* | |
|---|---|---|---|---|
| Potassium Perchlorate | 64 | | Potassium Perchlorate | 64 |
| Aluminum "Bright" | 14 | | Aluminum "Bright" | 24 |
| Aluminum Flitter 30/80 | 14 | | Aluminum Dark Pyro | 4 |
| Shellac 60 mesh | 8 | | Shellac 60 mesh | 8 |

damp these stars with a 10% solution of shellac in alcohol

| *Green/Silver* | | | *Red/Silver* | |
|---|---|---|---|---|
| Barium Chlorate | 25 | | Potassium Perchlorate | 70 |
| Barium Nitrate | 25 | | Strontium Carbonate | 12 |
| Aluminum "Bright" | 19 | | Accaroid Resin | 6 |
| Potassium Chlorate | 13 | | Aluminum "Bright" | 6 |
| Accaroid Resin | 7 | | Aluminum Flitter 30/80 | 6 |
| Dextrin | 5 | | Add alcohol | |
| Barium Carbonate | 4 | | | |
| Charcoal 150 mesh | 2 | | | |
| add water | | | | |

Purple stars can be made by substituting about half the Paris Green of the blue star composition with strontium carbonate.

### Yellow Illuminating

| | |
|---|---|
| Barium Nitrate | 68 |
| Cryolite | 10 |
| Aluminum Dark Pyro | 11 |
| Sulphur | 5 |
| Accaroid Resin | 5 |
| Boric Acid | 1 |
| add alcohol | |

*Pressed Stars*

Some stars are pressed dry in a hydraulic press or they can be pressed on a modified tablet press. Roman candle stars can be made by the latter method, but the scope is limited to certain gunpowder type mixtures. Chlorate stars are very sensitive for pressing in this way and usually become so hard when they are pressed that they will not ignite. Gunpowder, potassium nitrate and sulphur mixtures press quite well, but charcoal is not easy to press and stars containing a large amount of charcoal tend to break up as they are ejected from the molds. Most of the pressed stars are magnesium stars used principally for signalling in Very cartridges. The magnesium composition is pressed dry in a sleeve of paper or aluminum which has to be supported in a mold. Nitrates mixed with magnesium and PVC can be pressed dry in this way, though there is a limited shelf life for these articles since the magnesium is uncoated. In some compositions the magnesium is coated with linseed oil or drying oil and allowed to stand for some hours, prior to mixing in the other ingredients. Polyester resin is also sometimes incorporated into the composition prior to pressing. All magnesium stars require an igniter composition when the stars are pressed and so it is common practice to press an increment of igniter composition on top of the illuminating composition and finally an increment of some type of blackpowder. Igniters are usually mixtures of silicon powder, potassium nitrate and some type of gunpowder, but mixtures of silicon, lead dioxide and cuprous oxide are also used. According to the BIOS reports (12) the following were used during the last war:

| Red | | Green | |
|---|---|---|---|
| Strontium Nitrate | 55 | Barium Nitrate | 55 |
| Magnesium | 28 | Chlorinated PVC | 29 |
| PVC | 17 | Magnesium | 16 |

| Illuminating | | Yellow | |
|---|---|---|---|
| Barium Nitrate | 55 | Sodium Nitrate | 55 1/2 |
| Potassium Nitrate | 10 | Magnesium | 17 |
| Aluminum Dark Pyro | 21 | PVC | 27 1/2 |
| Sulphur | 8 | | |
| Barium Fluoride | 6 | | |

### Igniter

| | |
|---|---|
| Meal Gunpowder | 50 |
| Potassium Nitrate | 16 |
| Barium Nitrate | 16 |
| Aluminum Dark Pyro | 10 |
| Sulphur | 8 |

The stars were pressed at about 10 tons on 24mm.

Magnesium grades 3 to 5 are used in these stars, but it is necessary to experiment with the grades of magnesium to get the correct burning time. The red and green stars also make good pill box stars if 5% accaroid resin is added to the mixture which is then damped with alcohol. English and American formulations normally incorporate a percentage of potassium perchlorate into the composition in order to obtain a higher candle power e.g. Formula 62 in Ellern (5).

| | |
|---|---|
| Magnesium | 30 |
| Strontium Nitrate | 42 |
| Potassium Perchlorate | 9 |
| PVC | 12 |
| Laminac (Polyester) | 7 |

Dr. Becher (10) quotes the following star composition which is of interest:

### White

| | |
|---|---|
| Barium Nitrate | 50 |
| Strontium Nitrate | 10 |
| Aluminum Dark Pyro | 25 |
| Sulphur | 8 |
| Meal Gunpowder | 7 |

### Spherical Stars

Again, Dr. Shimizu has described the technique for the manufacture of these items on page 252. For many years it has been a mainly Japanese practice in order to obtain color-changing stars in their chrysanthemum bombs, but some European manufacturers are now beginning to make these stars by using a "Sweetie Barrel" similar to the photograph in Davis' book on page 291 (7).

# Chapter 7

# Colored Fires, Bengals, Lances, Portfires, Torches

Colored flames are made with almost the same compositions as those used for stars, except that fuels which are used to make stars burn rapidly (charcoal for example) are usually omitted. On the contrary it is usual in these items to add materials which will retard the burning rate. Colored fires, torches and bengal illuminations which are sold to the public are usually required to burn quite slowly, and normally have the burning rate reduced by adjusting the fuel. The use of coarsely ground materials, or the cutting down or increasing of the amount of fuel can have this effect, but it is also usual to add retardants such as sawdust, woodflour, starch or flour. The addition of these materials also makes the composition cheaper of course, but care must be taken with materials such as woodmeal since they can contain a good deal of water.

According to the older text books, extensive use seems to have been made of tableau fires in the past. This is not very common now, mainly perhaps because piles of loose composition burn very rapidly and the same effect can be obtained from a smaller quantity of material in the appropriate container. Nevertheless it has been a feature of English firework displays to burn heaps of red or green fire behind bushes. Smoke enhances colored flames since it tends to reflect light, and unless the colored fire is to be used indoors, no attempt should be made to cut out all the smoke. Some who have tried to do this have discovered that part of the effect was also sacrificed at the same time.

Weingart and others have set out compositions containing picric acid in color production, but this is highly dangerous with chlorates, quite pointless and expensive.

Containers for colored lights are normally made with a thin wall so that the tube will burn away with the composition in order to gain the best color effects. With these slow-burning compositions, the tube need only be dry-rolled and pasted at the edge; lances for instance are made in this way, but if the filling operation exerts any pressure it may be necessary to use a wet rolled tube; wet rolled tubes with thin walls can be quite strong. When thick walled tubes are used for colored fires they must be quite short (e.g. two to three inches long,) or else it will be found that the color deteriorates as the fire moves down the tube. The color is also destroyed by the yellow flame color of the burning tube. The best color effects are thus obtained with wide bore tubes of 3/4″ in diameter and upwards, and with thin walls.

It is well known that the human perception of color can be erratic and care must be taken when colors are produced in variety at close range. The human eye is more perceptive to reds than to greens for instance, a fact well illustrated when color changing lances change from green to red for then the change-over is easily noticed. Yet, if the color order is reversed with a change from red to green, the green cannot be seen until it has been burning for some time. In addition a red flame appears to be orange or yellow if it is observed for a period of time at close range. For these reasons it is best to arrange colored lights so that the observers see the reflected light but not the source itself. Aerial fireworks utilizing greater distances do not present the same problem though care must be taken with magnesium colors, for these do not always mix well in close formation. Red and white, green and white and possibly red and green magnesium colors will mix, but yellow must

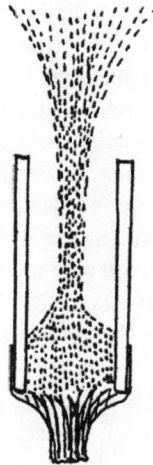

Fig. 28   Composition being poured into a tube

be used on is own, and to mix all four magnesium colors together would destroy the colors. Blue magnesium colors are not reliable, for the combustion temperature and the magnesium destroy the color, quite apart from the fact that the usual blue chemical combinations with magnesium are unstable.

### Colored Fires

The following compositions may be burnt as heaps of loose composition or gently tamped into short squat tubes. As it has already been stated, any length of tube can be used, provided that the tube will burn away with the composition; in this case the wall thickness of the tube would be about 1/16″. A short squat tube should be 3/4″ inside diameter, 3″ long with a 3/16″ wall. The composition should be charged into the tube with a drift and hand pressure only, or a funnel and wire could be used. Many of these mixtures burn more easily and reliably when they are gently compacted or even just shaken down. (Fig. 28)

|  | *Blue* | | | *Green* | | |
|---|---|---|---|---|---|---|
|  | A | B | C | A | B | C |
| Ammonium perchlorate | 46 | 30 | — | — | 50 | — |
| Potassium perchlorate | 26 | 40 | 75 | — | — | — |
| Basic copper carbonate | 10 | 15 | — | — | — | — |
| Accaroid resin | 15 | 15 | 12 1/2 | 12 | 15 | 20 |
| Stearine | 3 | — | — | — | — | — |
| Copper oxychloride | — | — | 12 1/2 | — | — | — |
| Barium Chlorate | — | — | — | 55 | — | — |
| Potassium chlorate | — | — | — | 33 | — | 26 |
| Barium nitrate | — | — | — | — | 35 | 54 |

|  | *Red* | | | *Yellow* | |
|---|---|---|---|---|---|
|  | A | B | C | A | B |
| Potassium perchlorate | — | 66 | 68 | 75 | 75 |
| Strontium nitrate | 66 | — | — | — | — |
| Strontium carbonate | — | 20 | 11 | — | — |
| Cryolite | — | — | — | 10 | — |
| Accaroid resin | — | 14 | 11 | 15 | — |
| Shellac 60 mesh | 17 | — | — | — | 15 |
| Woodmeal | — | — | 10 | — | — |
| Sodium Oxalate | — | — | — | — | 10 |
| Potassium chlorate | 13 | — | — | — | — |
| Charcoal 150 mesh | 4 | — | — | — | — |

|                          | White | | Silver |
|                          | A | B | A |
|--------------------------|----|----|----|
| Potassium nitrate        | 74 | 13 | 15 |
| Sulphur                  | 8  | —  | —  |
| Orpiment                 | 18 | —  | —  |
| Barium nitrate           | —  | —  | 45 |
| Potassium perchlorate    | —  | 64 | —  |
| Antimony powder          | —  | 13 | —  |
| Aluminum dark pyro       | —  | —  | 15 |
| Aluminum 30/80 flitter   | —  | —  | 20 |
| Accaroid resin           | —  | —  | 5  |
| Copal gum                | —  | 10 | —  |

*Bengal Illuminations*

This special type of flare candle is used for illuminating public buildings and is extensively used in Europe for castles which particularly lend themselves to this type of illumination. In order to fulfil the necessary conditions, the composition should produce maximum color, burn efficiently but as slowly as possible, (i.e. 40 to 60 seconds per inch), and not emit too much smoke.

The candles are made in various sizes, but the large ones are about 2″ in diameter, 12″ long and have a wall thickness of about 1/16″. The end of the candle is fitted with a wooden plug and a screw eye for attachment to the holder. The candle is fixed so that it burns in a horizontal position so that the dross will not run down the side of the

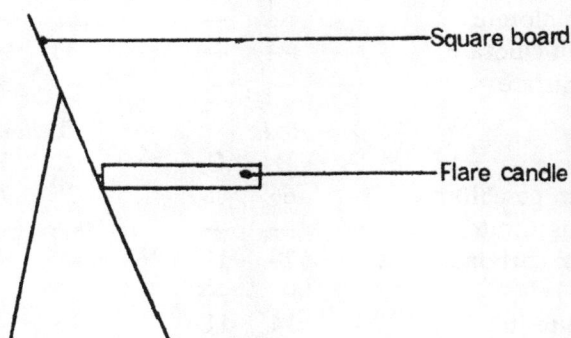

Fig. 29  Bengal illuminations

thin wall and accelerate the burning time. Lastly the candle is placed behind some kind of wall so that the onlookers do not see it burning. (Fig. 29)

The following compositions are typical. Paraffin oil is added sometimes as a retardant, and to give some protection against moisture which can be a problem in strontium nitrate mixtures which are not heavily compressed.

|  | Red | | Green | |
| --- | --- | --- | --- | --- |
|  | A | B | A | B |
| Strontium nitrate | 65 | 63 1/2 | — | — |
| Barium nitrate | — | — | 68 1/2 | 70 |
| Potassium chlorate | 20 | — | — | 16 |
| Potassium perchlorate | — | 16 | 15 | — |
| Shellac 30/200 mesh | 15 | — | — | — |
| Accaroid resin | — | 9 1/2 | 15 | 13 |
| Sawdust | — | 9 1/2 | — | — |
| Lampblack | — | 1 1/2 | — | — |
| Antimony metal powder | — | — | 1 1/2 | — |
| Paraffin oil | — | — | — | 1 |
| Approximate burning time in sec/inch | 45 | 30 | 20 | 40 |

It is always the case with firework manufacture that there are very many snags which practical workers discover only by experience. Strontium nitrate mixtures sometimes shrink if they contain water which subsequently dries out, leaving a gap between the composition and the tube. This naturally increases the burning time.

*Lances*

Large quantities of these items are manufactured every year for use in making set pieces, as fire pictures, mottos, motifs, all made with these little tubes of colored fire. Lances are usually 5/16" or 3/8" in diameter and about 4" long. The paper is usually an 0.004 bond or poster paper cut up into pieces 3 3/4" by 5", and they are dry-rolled on a former which is countersunk at one end. The papers are placed on top of each other with just the outside edge and side uncovered. A pasted sheet of paper is placed on the bench and the former is laid along the paper about 1/4" from the end. The paper is brought up and over the former and then rolled up to make the tube. The pasted edge which is left overhanging the end of the former is then pressed into the countersunk base of the former. This completely closes the base of

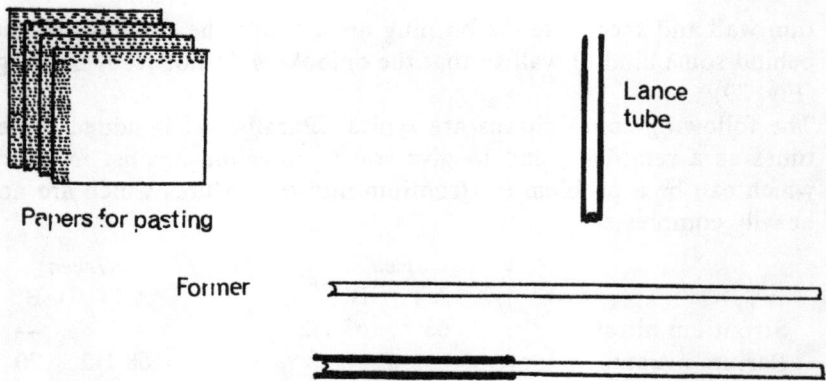

Papers for pasting

Former

Lance tube

**Fig. 30**   Apparatus for manufacturing lance tubes

the lance in a neat fashion, thus preventing the powder from trickling out. The following diagrams in Fig. 30 will help to make this clear. The following compositions are typical of those used for lances:-

|                        | *Red* | *Green* | *Blue* | *Amber* | *White* |
|------------------------|-------|---------|--------|---------|---------|
| Potassium perchlorate  | 70    | 30      | 64     | 73      | —       |
| Barium chlorate        | —     | 60      | —      | —       | —       |
| Accaroid resin         | 12    | —       | —      | —       | —       |
| Shellac 60 mesh        | —     | —       | —      | 15      | —       |
| Copal gum              | —     | 10      | 4      | —       | —       |
| Paris Green            | —     | —       | 32     | —       | —       |
| Strontium carbonate    | 18    | —       | —      | —       | —       |
| Sodium oxalate         | —     | —       | —      | 10      | —       |
| Charcoal 150 mesh      | —     | —       | —      | 2       | —       |
| Potassium nitrate      | —     | —       | —      | —       | 65      |
| Sulphur                | —     | —       | —      | —       | 20      |
| Antimony metal powder  | —     | —       | —      | —       | 10      |
| Meal gunpowder         | —     | —       | —      | —       | 5       |

A funnel and wire is used to charge the composition, but the base of the funnel has a nozzle which goes about half an inch into the lance tube. In this way, the lances can be filled to within half an inch of the top. The next operation is to add the topping composition in a similar way, to a depth of 5/16″. The old manufacturers used to use the white lance composition as a topping, but it is not wise to funnel and wire a sulphur mixture on to a chlorate or perchlorate and it would be very hazardous with barium chlorate. A simple mixture of potassium nitrate and accaroid resin serves the purpose quite well. The lance is finally primed with a paste of mealed gunpowder, dextrin and water.

As Dr. Shimizu points out on page 212, lances are sometimes a problem mainly because the flame is disturbed by the dross or they produce a long chimney-like ash which obscures the color and the brightness. In view of this, a good lance can only be tested at about 50 yards distance when it should be observed burning in a horizontal position. A lance 3 1/2″ long should burn about 50 to 60 seconds. Lances made with potassium perchlorate are less likely to chimney than those made with the chlorate, and some care needs to be taken with the choice of paper. Spirally wound lance tubes have appeared on the scene now but they are greatly inferior to the hand rolled variety.

*Portfires*

Portfires, which are used for lighting fireworks, are made in exactly the same way as lances, except that the tube is usually longer and has a slightly thicker tube. Portfires are normally 3/8″ inside diameter and about 15″ long. The paper for the 15″ size would be 7″ × 15″. Portfires usually burn three or four minutes, but they must not burn too slowly or they are apt to go out. The following compositions are satisfactory:-

|  | A | B |
|---|---|---|
| Potassium nitrate | 60 | 63 |
| Sulphur | 20 | 16 |
| Meal gunpowder | 20 | 11 |
| Antimony sulphide | — | 10 |

*Blue Lights Star Lights*

These little items seem to have almost disappeared from the market now. In effect they were small lances 3/16″ inside diameter, 6″ long with a wall thickness of 1/16″. The compositions were quite slow burning and produced either golden spur fire or molten sparkling drops of the type mentioned by Dr. Shimizu on page 226. The following compositions were used:

|  | *Blue Light* | | *Star Light* |
|---|---|---|---|
|  | A | B |  |
| Meal gunpowder | 51 | 13 | 6 |
| Potassium nitrate | 35 | 62 | 62 |
| Sulphur | 14 | 19 | 19 |
| Orpiment | — | 6 | 9 1/2 |
| Lampblack | — | — | 3 1/2 |

The potassium nitrate was sometimes a mixture of fine powder and the more coarse crystalline powder. These fine tubes were not always charged with a funnel and wire; the composition was merely poured over the open ends of the tubes and shaken down in a similar way to the method used for filling English-crackers (see page 167). Again, not surprisingly, many compostions filled in this loose manner function better than those which are carefully charged.

*Torches*

Weingart wrote a great deal about these items but they are seldom used now in Europe. The reason for this could be that they are comparatively expensive, or that they are liable to deposit dross, make a lot of smoke, and are liable to cause accidents in crowds. The most commonly used torches today are made of a piece of Hessian or sackcloth which is rolled around a former and then dipped in molten paraffin wax. It is quickly removed from the former and subsequently fitted with a suitable handle. These torches are non-pyrotechnic, but are cheaper and safer. Pyrotechnic parade torches and railway fusees are 3/4″ to 1″ in diameter and 8 to 18″ long. The tubes are normally wet rolled with a piece of thin kraft paper which makes about four to five turns on the former. Lance compositions are quite suitable for colored torches and, if need be, these can be made cheaper by the addition of 5% to 10% woodmeal or 5% starch or wheatflour. The following compositions are also quite useful:-

|                              | A  | B  | C  |
|------------------------------|----|----|----|
| Potassium perchlorate        | 52 | —  | 25 |
| Strontium nitrate            | —  | 45 | —  |
| Aluminum 150 mesh "Bright"   | 24 | —  | —  |
| Aluminum Flitter 30/80 mesh  | 20 | —  | —  |
| Aluminum dark pyro           | —  | 18 | —  |
| Dextrin                      | 4  | —  | —  |
| Barium nitrate               | —  | 76 | —  |
| Sulphur                      | —  | 4  | —  |
| Petroleum jelly              | —  | 2  | —  |
| Shellac 30/200 mesh          | —  | —  | 14 |
| PVC                          | —  | —  | 6  |
| Linseed oil                  | —  | —  | 1  |
| Magnesium Grade 0            | —  | —  | 9  |

Parade torches are usually furnished with a wooden handle which is glued into one end, while the other end is primed with a modified gun-

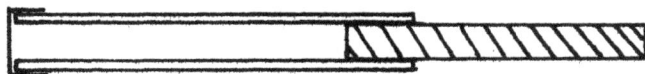

**Fig. 31**   Torch or hand flare

powder and touchpapered. (Fig. 31) Signal torches are frequently primed with a mixture of potassium chlorate and fine charcoal which has been mixed into a paste with dextrin and water. A small wooden striker on which is pasted a slurry of red phosphorus is also provided for ignition. More sophisticated signals have striker ignitions embedded in polythene units, ignition being achieved by pulling the wire to release the striker.

*Flares*

In some circumstances, illumination can be achieved almost as well with aluminum as with magnesium. The following compositions can be charged into wide bore tubes with thin walls, but care must be taken when charging mixtures of barium nitrate and aluminum, for these mixtures are more sensitive than one might think. It is in fact much wiser to press them into tubes on an Arbor press.

|                        | *A* | B  | C  | D      | E  |
|------------------------|-----|----|----|--------|----|
| Barium nitrate         | 56  | —  | —  | 22 1/2 | 50 |
| Strontium nitrate      | —   | 60 | 42 | —      | —  |
| PVC                    | 21  | 15 | 12 | 13     | —  |
| Magnesium grade 0      | —   | —  | 30 | 35     | —  |
| Potassium perchlorate  | —   | —  | 9  | 22 1/2 | —  |
| Polyester              | —   | 5  | 7  | 5      | —  |
| Magnesium grade 4 or 5 | 16  | 20 | —  | —      | —  |
| Aluminum dark pyro     | —   | —  | —  | —      | 35 |
| Potassium nitrate      | —   | —  | —  | —      | 15 |
| Montan wax powder      | 7   | —  | —  | —      | —  |

Formulas A, C and D are from **BIOS** (12) sources and Ellern (5), the two latter being from the American patent literature.

Compositions containing barium nitrate and aluminum are also somewhat difficult to ignite and it is always necessary to add a suitable ignition mixture such as those mentioned on page 93 or a hot burning antimony white star mixture would work equally well.

Magnesium flares are rather more complex to manufacture, mainly because they depend so much on the correct preparation of the materials and they require pressing under heavy pressing loads. Ellern and others have printed many typical formulations.

*Waterfalls*

In view of the fact that waterfall compositions are closely related to the rest of the fireworks described in this chapter, they will be described here.

The general principle of formulation is to ensure that there is an excess of aluminum present, in order that the burning material will fall to the ground with the appropriate sparks. The tube is usually a wet rolled tube consisting of four or five turns of an 0.010 kraft paper on a 3/4" or 1" former. The tube is six to nine inches long and naturally must burn away with the composition or else the essential dross will not fall to the ground.

Old formulations were made with potassium chlorate e.g.

| | |
|---|---|
| Potassium chlorate | 72 |
| Aluminum "Bright" | 28 |

This was damped with a 10% solution of shellac in alcohol, mainly to control the aluminum dust, but the small amount of well distributed fuel also improves the burning and holds the composition together. Naturally it is much safer to use potassium perchlorate, and in this case a slightly larger proportion of aluminum is required.

| | |
|---|---|
| Potassium perchlorate | 50 |
| Aluminum "Bright" | 25 |
| Aluminum Flitter 30/80 mesh | 12 1/2 |
| Aluminum Flitter 5/30 mesh | 12 1/2 |

The usual method of charging is to cover one end of the tube with a thin paper "drum head" or a piece of muslin. A small quantity of igniter composition is placed in the tube first, followed by successive increments of composition which are lightly tapped down with a mallet and drift. Bundles of tubes with a maximum of 5 lb. of composition are damped at one time. The tube is finally closed with a disc.

Many people prefer to use barium nitrate as an oxidizer, and while this is safer it is also inferior, for there is a tendency to get a large glare at the mouth of the tube and a weaker drop effect. Dr. Becher (10) gives the following composition which is fairly typical:-

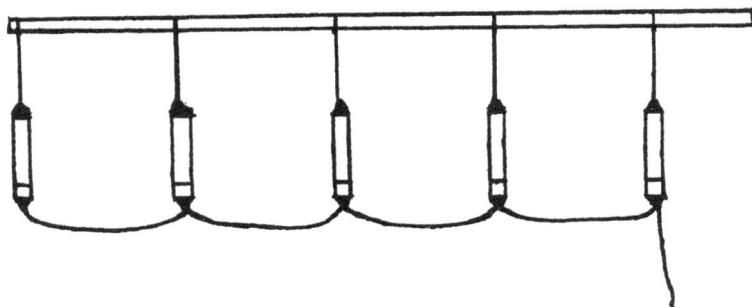

**Fig. 32**  Waterfall

| Barium nitrate | 52 |
| Potassium nitrate | 8 |
| Aluminum dark pyro | 16 |
| Aluminum flitter | 21 |
| Meal gunpowder | 3 |

In this particular instance the increments of composition should be pressed dry with a hand press and not charged with a mallet. The igniter composition is also quite essential with this particular mixture. It is also necessary to adjust the aluminum to fit the particular grades used; a bright powder can be used equally well instead of pyro, but this will necessitate the adjustment of the flitter.

All the shower sticks are finally primed on the outside of the drum head with mealpowder, capped and prepared for use as waterfalls. Some people insert a piece of wooden dowel into the end of the shower stick and then nail this on to a length of board which is fixed so that the shower sticks burn in a horizontal position. In England it is common to fasten a piece of thick string on the end of the shower and use this to fasten the showers on to a board. In this way the showers actually move during burning. (Fig. 32)

Good waterfall effects can also be obtained by using a bank of iron or titanium gerbs of 1″ bore, fixed to burn in a horizontal position.

# Chapter 8

# Roman Candles, Comets, Mines

*Roman Candles*

There can be no doubt that it requires much practice to make really good roman candles, and the best ones are still made by hand. So many factors play a part in the performance of the candle with the inevitable result that there is always something to go wrong. Each of the factors will therefore be examined in turn and the possible snags pointed out.

Firstly, the tube is all important, and for the purposes of this essay, only one size of roman candle will be referred to, the tube being 5/8" inside diameter, at least 12" long and having a wall thickness of about 1/4". The wall needs to be of this thickness or else the tube will catch fire during the performance of the candle. The best tubes are hand rolled and it is quite essential that the inside lap of paper is completely stuck down or else the powder will creep down the loose paper, cause a "blow-through" and eject the stars like machine-gun bullets. The stars are made in a 9/16" former which means that they just comfortably slide down the inside of the tube. In addition, as would be expected, stars made in a 9/16" former shrink very slightly during the drying process, just as a wet rolled tube on a 5/8" former also shrinks during the drying process, with the result that both factors might play a part. Stars which are too tight are ejected so fast that they are frequently extinguished because they are projected too rapidly. Similarly, stars which are too slack often land on the floor still burning, since they were projected only a few inches into the air.

The quality of the paper of the tube is also quite important. A good quality kraft paper for example, rolled with a good casein glue, provides a tough tube which hardly burns away during the firing of the candle. If a good quality grain powder were used for such a tube, the stars would be ejected with increasing force as the fire proceeded down the tube and would probably be extinguished. The reason for this is not far to seek because the tube is not burning away at the top and a high pressure is thus maintained. On the other hand, the use of a cheap paper could mean that the tube burns away at the top, allowing the gas to escape, with the result that the stars are not projected quickly enough. Clearly the manufacturer uses a good quality grain for poorer quality papers and vice versa. In practice it is usual to use the cheaper paper. The 9/16″ stars are usually about the same length, but the burning rate of the composition depends on the way the stars are placed in the tube, for there are two methods. The better method uses very fast burning stars such as the ones on page 87 and surrounds each star with a small amount of gunpowder to make sure that it will be ejected immediately. Since the star is ejected so rapidly it is essential that it should be fast burning or else it will be extinguished. The second technique places no powder around the star but uses a much slower burning star which actually burns for a second or two in the tube before it is ejected. Gas is lost up the side of the star, which is not ejected as high as those made by the first method. The latter method also produces candles which are more erratic in their performance, for a good roman candle should eject all its stars regularly to the same height, the stars burning out just as they turn over to come down again.

The roman candle delay also plays a part, for its burning characteristics are directly related to the grain powder charges under each star. Manufacturers naturally tend to have their personal preferences, some maintaining that it is essential to granulate the composition by damping it with water and subsequently drying it. If this granulation takes place for a definite purpose, such as making the composition dust free and therefore cleaner to handle, or easier to manipulate in the measuring boards for mass filling, this is fine, but it is otherwise unnecessary. Some argue that the fuse should contain a large proportion of gunpowder or sulphur or both, though it is not always clear why this view is held. The most significant factor would appear to be the need of a reasonably dense fuse which consolidates well, a condition mainly satisfied by the presence of sulphur, potassium nitrate, gunpowder and not too much charcoal. The fuse ideally should produce a good show of golden sparks but this is not necessary of course and there is always a tendency to produce too much smoke as would be expected.

The following composition is quite a satisfactory one, since it combines most of the characteristics mentioned:

| | |
|---|---|
| Potassium Nitrate | 50 |
| Meal gunpowder | 22 |

> Charcoal 40/100 mesh      11
> Charcoal 30/60 mesh       11
> Sulphur                            6

The manufacture of a 12″ roman candle, for example, would take place as follows. The tube, (see the dimensions at the beginning of the chapter) is placed on a small nipple and a charge of clay is placed inside. Using a drift which slides easily down the tube, the clay is firmly consolidated. A charge of English F or FFF grain powder is next placed in the tube, followed by a star. A small charge of gunpowder or fine grain powder is then placed on top of the star and the tube is tilted backwards and forwards so that the powder will trickle down the side of the star. A scoopful of delay fuse is now placed in the tube and this is consolidated lightly with a mallet, and it should be noted that this is where the skill really lies. If the delay is rammed too lightly or too firmly the candle will not function properly, with the result that the stars are ejected too fiercely or in quick succession. Only experience will teach the operator just how hard to strike the drift.

The fuse and clay scoops are usually semi-cylindrical, being beaten out of a sheet of copper. Grain scoops are smaller, but are made in a similar way. A 5/8 by 12″ roman candle takes about six stars. The quantities are roughly as follows:-

> Fuse scoop holds approximately      4 gm.
> Grain scoops hold 0.45, 0.55, 0.60, 0.90, 1.20, 1.50 gms.
> About 0.50 gm. of gunpowder is placed on each star.

The smallest grain charge goes in the bottom of the tube on top of the clay and then the charge increases as it gets nearer the top. Fig. 33 shows a roman candle in cross-section.

In practice, these candles are not filled singly of course. Display candles are usually filled in small bundles, while commercial items may be charged a hundred or more at a time. American practice appears commonly to employ gang rammers, but this is unusual in Europe where the candles are rammed singly or six at a time. The European method

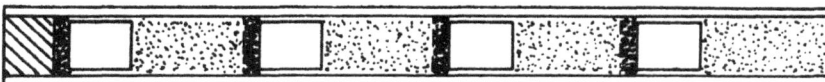

**Fig. 33**   Roman candle

**Fig. 34**   Mass charging of Roman candles.
(Courtesy Pains-Wessex, Ltd.)

no doubt springs from a fear of the mixture of chlorate in the stars and the sulphur in the fuse plus friction from the gang rammers. Fires and explosions have certainly been caused by gang rammers. Figure 34 shows roman candles being charged at Pains-Wessex Ltd.,

In mass-filling the grain and the fuse are added by means of the measuring boards described by Weingart. Thick boards made of wood or synthetic resin are drilled with suitable holes just large enough to admit the required quantity of fuse or grain. Another similar board is then placed underneath with corresponding holes which will allow the powder to trickle through into the tubes. If the two boards are arranged so that they will slide over each other, the powder can be contained in the holes of the top board or allowed to run through into the tube by merely sliding one board over the other. The principle is not unlike the one used by organ builders for making the wind available to ranks of pipes. The diagrams will explain the method more easily than a detailed description, and greater detail is pointless since every manufacturer will modify the method to suit his own purposes. (Fig. 35).

A wide variety of roman candles are made, with inside diameters as small as 3/8″ and as large as 2″. Most of the candles eject stars such as colors, glitters, aluminum and lampblack stars, but more sophisticated

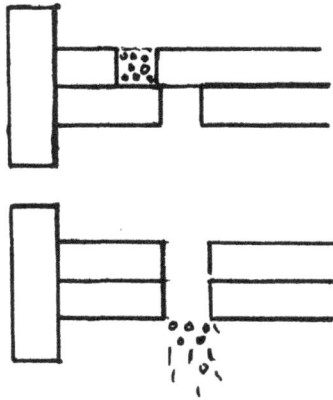

Fig. 35    Sliding boards for charging powder

versions eject pressed magnesium stars, picrate whistles, humming stars, flash and sound, serpents and large comets. The special larger bore candles usually have to be filled singly and require great care in their manufacture since they are such expensive items.

*Comets*

Comets are single roman candle stars in their simplest form, but they are usually gunpowder type compositions since they are ejected at great speeds. The compositions are damped with dextrin and water or accaroid resin and spirit and then consolidated in a single star pump. Large star pumps are made of brass or a mixture of brass and copper and so arranged that they can be charged with a mallet or placed in a press. The smaller comets of about 1″ in diameter can be adequately charged with a mallet, but the larger comets should be pressed in order to get greater consistency. When comets are being manufactured, it is usual to press more composition than is necessary, eject the excess and then cut off this excess with a knife. (Fig. 36)

The majority of the smaller comets are placed in cylinder bombs but they can also be effectively used when fired from the ground in large numbers in quick succession. Comet batteries consist of steel plates on which are welded a dozen or more steel tubes about 8″ or 9″ long. Holes are drilled in the base of each tube to allow a piece of slow fuse to be threaded through each one. A charge of grain powder, possibly about 10 gms, is loaded into each of the tubes and then a comet star is placed on top of the charge. Lastly a tight fitting wad is rammed on top of the star. The "wad" need only be a sheet of paper which can be

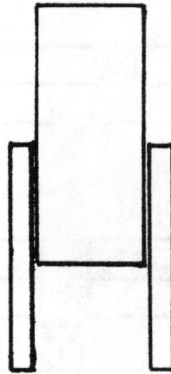

**Fig. 36**  Pump for comet stars

consolidated with a wooden drift. Naturally care must be taken not to ram the tube too tightly or else there is a danger of bursting the tube. Large bore tubes should never be rammed in this way. (Fig. 37).

Larger comets of 1-1/2" or 2" in diameter are either solid or filled with colored stars. The star is formed with a chamber in one end of the star so that a colored star or stars can be placed into the cavity which is formed. (Fig. 38).

The stars are finally covered with a pasted paper coat. The paper covers one end and three quarters of the length of the star. Solid stars are not normally covered. Crossette stars or splitting comets usually have a small charge of flash powder instead of colored stars. These stars are very effective as the streaming comet suddenly bursts into many fragments.

Effective comet bombs can be made by pressing a star in the bottom of a paper tube and then filling the remainder of the tube with stars or

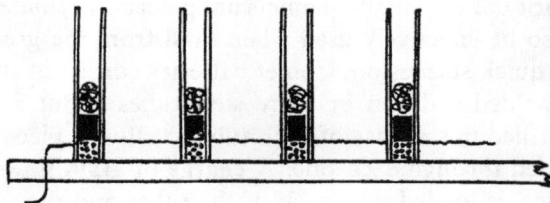

**Fig. 37**  Comet battery

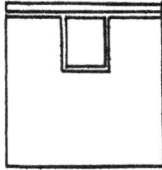

**Fig. 38**  Comet star

other effects. The result is a spiral of fire terminating in a star burst. These items need to be carefully manufactured otherwise they are liable to "blow through" in the mortar. (Fig. 39).

*Mines*
The ordinary mines are simply a charge of gunpowder and stars at the base of a strong tube which acts as a gun. The "Bags" of commerce, or Pot à Feu as they are sometimes called, consist of about three turns of a medium kraft paper dry rolled on a former which is about 1/2″ less than the diameter of the mortar. A 4 1/2″ bag for example is 4″ in diameter and about 6″ high. The paper for this bag would be about 40″ long for the rollup and about 12″ wide. A gunpowder charge of about 50 gm. of grain powder is placed in a small bag which is then

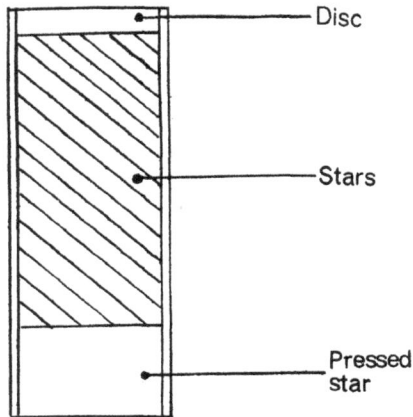

**Fig. 39**  Small comet bomb

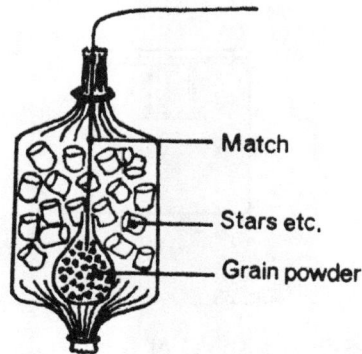

**Fig. 40**  A "bag" or mortar mine

tied to the end of a length of piped match. This bag is then placed into the bottom of the main bag and stars, tourbillions, flash charges, comets etc. are then loaded on top of the powder charge to a depth of about 6″, and the bag is tied off above the stars. Figure 40 will explain the method more easily.

Commercial mines are usually made in wide bore tubes about twice the height of the diameter. The tubes are fitted with a strong base of paper or wood which is securely glued into place, and a weakly fitting

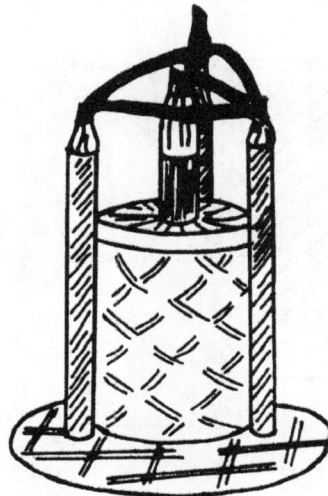

**Fig. 41**  Devil among the Tailors

lid which will blow off easily. Sizes vary from quite small tubes 1″ in diameter up to quite large ones 5″ or 6″ in diameter. Mines of this type are normally fitted with a central delay fuse which consists of a long narrow tube which contains no clay at the base, but is merely primed so that the fire can transfer to the powder charge. Roman candles are also used as delays. Special mines called "Devil among the Tailors" consist of a mine with a central roman candle and three other candles fixed to the outside of the tube. The device thus produces a fine display of colored stars which terminate in a mine burst of crackers, stars, serpents. (Fig. 41).

# Chapter 9

# Noisemakers

### Flash and Sound

Explosive fireworks more than anything else have been the cause of serious accidents and they have probably done more damage to the firework industry than anything else! Flash crackers in their various forms were on sale to the public for far too long and happily have been forbidden in many European countries. Gunpowder "bangers" are still sold to the public but they are mostly of interest only to schoolboys who frequently make a nuisance of themselves by throwing them at people.

It is also a tragic fact that many young people cannot resist the urge to create explosions, and in their enthusiasm they choose unstable and highly sensitive mixtures which they charge into unsuitable containers in highly dangerous ways. Fatalities caused by sodium chlorate mixtures being charged into containers made of metal (of all things!) must be legion.

There can be no doubt that even among firework manufacturers, compositions vary very much in their sensitivity, but so much depends on what is required of the mixture.

Flash compositions of some types will cause explosions in the lightest of containers - even sometimes in just a few turns of paper - but it also happens that these compositions are extremely brisant and sensitive. Horrifying mixtures of potassium chlorate, pyro aluminum, sulphur and barium nitrate have been employed and should be avoided at all costs. Mixtures of the perchlorate, sulphur and bright aluminum are safer and appear to be used extensively in the U.S.A. and Japan (page 231) but even these would be considered dangerous by many of us in Europe. In fact the more common European technique is to use a strong paper tube with a composition consisting simply of potassium perchlorate and dark pyro aluminum e.g.

|                        | A  | B  |
|------------------------|----|----|
| Potassium perchlorate  | 66 | 70 |
| Pyro aluminum          | 34 | 30 |

Sometimes barium nitrate is used as the oxidizer e.g.

|                | |
|----------------|----|
| Barium nitrate | 68 |
| Pyro aluminum  | 23 |
| Sulphur        | 9  |

These mixtures are not quite so easy to ignite, but normally do so with Bickford fuse and they easily ignite with quickmatch and gunpowder primes.

It must also be pointed out that dark pyro aluminum is a dark grey impalpable powder with no visible silver particles, in fact it does not even look like aluminum powder. Very many so called pyro powders are much coarser than the European dark pyro powders.

The explosions are normally made out of a paper tube 1/16 of an inch thick and upwards which is closed at each end with clay or corks. A piece of Bickford or other fuse penetrates the tube in some manner and is firmly glued in position. The containers are filled about one half or two thirds full of flash powder which is charged loosely and not compressed. Tubes made of metals or brittle materials should never be used because of the danger of flying debris. The barium nitrate mixture is less fierce than the perchlorate composition and in this case the tubes are filled quite full. Magnesium compositions are also sometimes used, but naturally they are less stable and rather more sensitive.

Maroons are sometimes filled with FFF grain powder but they need strong tubes. In former times, and occasionally still, the cores were wound with strong twine and then glued on the outside.

Photoflash mixtures made with potassium perchlorate, barium nitrate and atomized aluminum have much slower burning speeds and require a very strong tube to produce a loud noise.

## Whistles

The oldest whistles were made with potassium picrate. Picrate whistles are very shrill and can be very entertaining in roman candles with their black tails.

Potassium picrate is not normally manufactured by the picric acid manufacturers since there is little use for this material, though the principle English manufacturer does in fact produce small quantities for the firework trade. Dr. Shimizu describes the manufacture on page 234.

Picric whistles are not popular with the firework makers mainly because no-one cares to work with them. The salt stains the fingers and clothing a bright yellow and the taste is bitter and unpleasant.

It is well known that picric acid and its salts are sensitive to impact and great care needs to be taken when charging potassium picrate. In fact it is much wiser to consolidate the material with a small hand press. Lead picrate is particularly sensitive; therefore lead covered surfaces must be avoided.

Strong narrow bore tubes e.g. 1/4″ inside diameter, 1 1/4″ long and with a 1/4″ wall can be charged quite full, though the material needs to be well consolidated or else it will explode. If neat potassium picrate is used without any additives, the tube must be full to the top and not primed. Priming causes explosions also.

Other types of whistles operate in such a manner that it is essential to have an empty space over the composition, since the empty tube acts as a kind of resonator. Potassium picrate, as it has already been stated, will work in short tubes without any additive, but in long tubes it must be diluted. Up to 20% of potassium nitrate or up to 15% of stearin or powdered asphalt can be added for this purpose.

Whistles should only be lightly capped with no more than two turns of thin kraft paper or they are liable to explode.

Picric whistles are suitable for rockets, roman candles and ground pieces, but they should never be placed in shells lest the lifting charge detonates the picrate.

Many people regard potassium picrate as an exceptionally hazardous material, but there can be no doubt that very many accidents have been caused with the principal alternative, a mixture of potassium chlorate and gallic acid.

A mixture of three parts of potassium chlorate to one part of gallic acid is normally used for these whistles, though it appears that American formulations include a small percentage of accaroid resin. Potassium perchlorate cannot be used with gallic acid. The mixture is extremely sensitive and the greatest care should be exercised during mixing and charging, for explosions have taken place during mechanical mixing and fires have occurred during sieving and charging. The mixture is charged into tubes from 1/4″ to 3/8″ in diameter, to a depth of 1″ in about four increments. The top 1 1/2″ of the tube should be left empty. Two pieces of quickmatch are inserted into the empty space before the tube is capped. An additional disadvantage of this mixture in the high price of gallic acid.

In recent years there has been a tendency to make whistles with a mixture consisting of 70% potassium perchlorate and 30% sodium salicylate. Again, in the U.S.A. they add a small percentage of accaroid resin, and this is not necessary. In fact this mixture is hygroscopic and whistles should not be kept for long periods, particularly in unheated magazines over the winter. Oddly enough the addition of accaroid resin to this mixture seems to increase the tendency to absorb water, but the reason for this is not clear. Polyester can be added to the salicylate composition but it does not give complete protection from moisture. Potassium benzoate can also be used in place of sodium salicylate. This type of whistle can be pressed or hand charged, but the potassium perchlorate and most of the sodium salicylate must pass a 120 mesh sieve. The exact particle size is something of a problem and batches of sodium salicylate vary somewhat in their performance.

The mechanism of the whistle is very interesting. The crystals decrepitate and decompose in an oscillating manner. The whistling sound is produced by the rhythmic acceleration and stopping of the reaction in the tube.

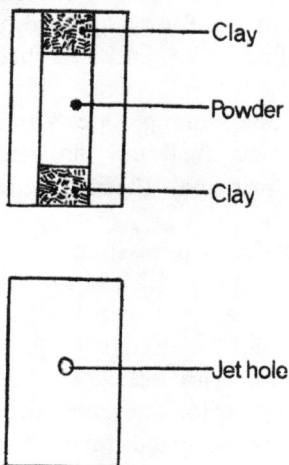

**Fig. 42**  Firework hummer

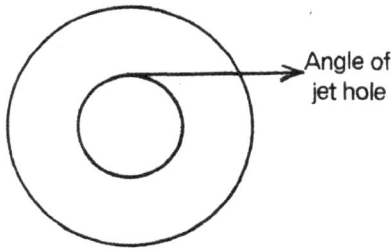

Angle of jet hole

**Fig. 43** Angle of the jet hole of a firework hummer

*Humming Fireworks*

Humming fireworks are constructed in a manner which will allow a jet of gas to issue out of a tube at such a speed and at such an angle that the tube will be caused to rotate on its own axis. In the air this causes a humming noise.

The smallest hummers are made out of narrow bore tubes about 5/16″ internal diameter and with walls 1/4″ thick. The tubes are about 1 1/4″ long and must be blocked with clay at each end. Inside the tube between the clay plugs a very fast burning mixture of gunpowder and aluminum is carefully charged. (Fig. 42)

The purpose of the aluminum is merely to slow down the meal-powder to prevent an explosion and also to produce some silver sparks. 7% to 10% of fine titanium or silicon can be used for the same purpose.

Larger hummers 5/8″ inside diameter often employ a little charcoal in addition to the gunpowder and aluminum and frequently have two holes.

The gas jet issues out of the tube from a small hole pierced through the wall in the centre. The size of the hole is, of course, related to the burning speed of the powder.

It will be observed from Figure 43 that the angle of the hole is such that the tube will be made to spin in flight. Needless to say it is better to pierce the hummer tube since drilling is rather hazardous. For ignition, a piece of quickmatch is placed in the pierced hole and then secured with a piece of tape. It is possible to drill the empty tubes first and then charge a piece of match into the composition during the filling operation.

# Chapter 10

# Rockets

## R. G. Hall

Rockets are one of the oldest pyrotechnic devices. Their history can be traced from the Middle East, where they were used as weapons, to India, and to China where, it is believed, they were first used for display purposes. The manufacture of rockets was well known in Europe during the early middle ages, and the origin of the word is thought to be based on the similarity in appearance of a rocket on its stick, to the round piece of wood used to cover the point of a lance in mock combat. This was known as a "rockette" an Italian diminutive of "rocca" a staff.

All rockets, from the smallest firework to the giant satellite carrying Saturn, have four basic components in common. They are:

1. A case or rocket motor
2. A "choke" or Venturi
3. A propellant charge
4. A flight stabilizing device

Rockets are reaction motors. On ignition, the propellant charge produces gas at a high temperature and so the internal pressure of the gases in the rocket body is raised. By allowing the gases to escape through a narrow opening, the pressure at this opening, or nozzle, falls and the velocity of the gases is increased. At any time, the momentum of the rocket must equal the momentum of the escaping gases so that, very simply, we have the following equation

Mass of rocket and unburnt propellant x The velocity of the rocket = Mass of escaping gas x the Velocity of the gas.

Rocket bodies or cases may be made from a variety of materials. Simple firework rocket bodies are almost exclusively rolled from a good quality Kraft paper. After cutting to size, the paper is pasted and tightly rolled on a cylindrical former and allowed to dry. It makes little difference whether this operation is done by hand or by machine, except that when the cases are machine rolled a slightly inferior quality of paper may be used, as the tighter machine rolling gives greater mechanical strength to the finished case. Care should be taken to ensure that the inside surface of the finished case is smooth and regular, and that the inside edge of the paper is firmly pasted down. This is important since it is possible to dislodge this edge during the charging operation and to turn it over into the composition. This would seriously affect the consolidation of the propellant charge and would most likely lead to the bursting of the rocket on ignition. The cases should be thoroughly dried. That is, they should be allowed to reach an even moisture content in equilibrium with the atmosphere in which they are to be charged. Failure to observe this simple precaution can cause swelling or shrinking of the case after charging and the subsequent malfunctioning of the finished rocket. For signal rockets, which have to stand more vigorous handling and storage conditions, the cases may be made from phenolic resin impregnated papers. After rolling the resin is cured or polymerized in a heated oven and this process

**Fig. 44a**  Rockets choked when damp

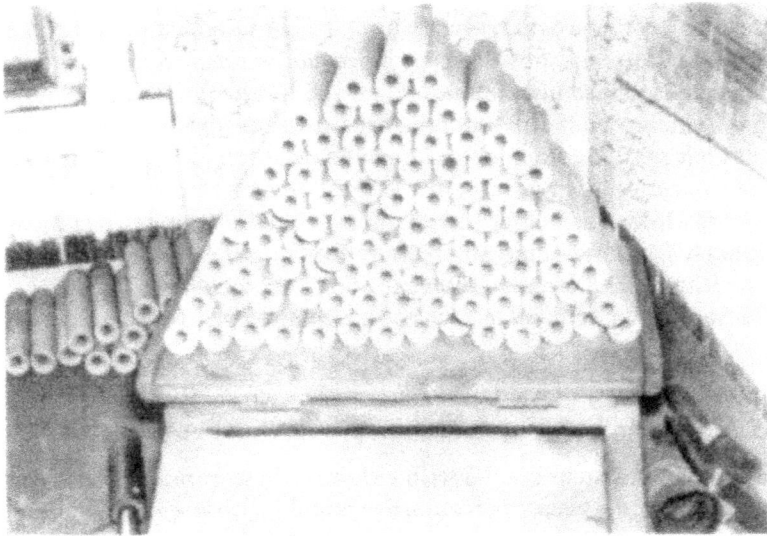

**Fig. 44b**  Rockets choked with clay. (Courtesy Brock's Fireworks.)

gives a rigid case of great mechanical strength. Metal cases of steel, or aluminum alloy, are also used and these are often varnished on the inside surfaces to give greater adhesion between the propellant and the case wall and sometimes to prevent undue heat transfer along the case. Good quality thin paper linings, which may or may not be varnished, are also used for the same purposes.

The "choke" or venturi is an important part of the rocket. The pressure energy of the propellant gases is converted to kinetic energy by this device. In firework rockets the "choke" is quite crude and is often formed by constricting the rolled case whilst it is still damp. The constriction, which may be from one to one and a half inches from the end of the case, is then tied with string to prevent its opening out on drying. A slightly more refined "choke" is formed by placing the dried case on a metal step which has a nipple of the required "choke" dimensions. Figs. 44a and 44b

A scoopful of dry clay is poured into the case and pressed or rammed into the case until it is solid. The case and its clay "choke" is then removed from the nipple. Nipple sizes and dimensions will be given later. In signal and line carrying rockets, the "choke" may be no more than a cylindrical hole bored in a metal plate which in turn is roll crimped or riveted to the rocket body. This is a common form of "choke" where black powder type propellants are used. When more sophisticated propellants are employed the nozzle and venturi should be so designed that the maximum thrust is given by the burning propellant charge. Economic factors and the desirability of mass production of nozzle units invariably force the designer to compromise and the result, while satisfactory, is not always the best which is technically possible. A full mathematical treatment of nozzle and venturi design is to be found in the many technical books on rocketry (22).

The traditional propellant charge for the firework rocket is either

1.  Gunpowder
2.  A mixture of mealed gunpowder and charcoal
3.  A simple mixture of potassium nitrate, sulphur and charcoal.

To obtain consistent results with either of these compositions, a few simple but very necessary precautions should be observed. When grain gunpowder is used alone, the grains should be small, regular and free from dust. They should be quite dry and show no signs of white crystals on the surface when examined under a powerful lens. The potassium nitrate and sulphur should be of a good commercial quality and the particle size should be closely controlled. Samples passing 120 British Standard Sieves and retained on 160 B.S.S are ideally suited for the mixed type of composition. The moisture content of the ingredients should be as low as possible and this level should be consistent. Any drying out of the composition after charging would invariably lead to cracking of the charge and to increased pressure build up and bursting of the rocket on firing. Charcoal is the most difficult ingredient to control. It is imperative that the charcoal be kept in a dry, warm store, if the regularity of the burning rate of the finished composition is to be maintained. The particle size should be sufficiently large to give a good glowing "tail" when the rocket is fired. A charcoal passing 40 B.S.S. and retained on 80 B.S.S. is useful for small rockets, and a charcoal containing 10% passing a 15 to 20 B.S.S. and retained on 40 B.S.S. gives a good display in larger rockets. The ash content of the charcoal should be consistent from batch to batch and the type of wood from which the charcoal is made should also be specified. Generally, hard wood charcoals give compositions which burn more slowly than similar compositions made from soft wood varieties.

When a particular formula has been found to give a satisfactory rocket subsequent batches of composition can be checked for consistency by column test. A weighed amount of composition is pressed in equal increments, into a gun metal tube about 3/8″ diameter until the column of composition is about 2″ in length. This length is controlled by a mark on the ram.

The measurements given are purely arbitrary and any reasonable measurements can be used but, once the weight of the composition, the number of increments, the diameter of the tube, and the length of the composition has been decided by the manufacturer, they should be strictly adhered to. The column of composition is fired by a piece of quick match and the burning time is checked by a stop watch. A variation in the burning time of $\pm$ 1.1/2% is allowable. The gun metal tube is washed in warm water, dried, and is ready for the next test. This test is quick, reliable, and easily carried out. Any discrepancy in mixing or in the purity of the ingredients shows up in increase or decrease of the burning time.

Modern propellants for signal rockets may be of the double or triple base nitro-cellulose powders. These are mixtures of nitro-cellulose with nitro-glycerine, or of nitro-cellulose with nitro-glycerine and nitro-guanidine. More usually plastic/perchlorate propellants are employed, since these allow the use of simply designed motors. These propellants are capable of being molded to any desired shape in order to give the correct burning characteristics. Typical formulations would contain polyisobutene, ammonium perchlorate, ammonium picrate (13). There are three basic types of solid propellant rockets. In the first type, the composition is pressed into the case in a solid mass and is ignited at one end only. The propellant is said to burn in a "cigarette" or "end" burning manner.

It will be seen that only a limited burning surface is available for gas production, and therefore, a fairly quick burning composition is needed in order to sustain an adequate internal gas pressure. In firework rockets, these compositions are usually compounded of mealed gunpowder and fine charcoal. The charcoal is usually present in proportions varying from 20% in rockets of 3/4″ bore to 10% in the smaller 1/2″ diameter rockets. Fine grained gunpowder is also used and such rockets are quite powerful for their size, but they lack a spectacular "tail" and in the writers opinion are not so interesting as those rockets containing charcoal. In these rockets the "choke" is of clay as already described and the diameter usually varies between 3/32″ and 3/16″ depending upon the diameter of the rocket and the type of composition used. This end burning type of rocket is usually pressed with the case supported in a metal mold in order to prevent collapse of the paper case. The composition is put into the case in four to six equal increments and each increment is pressed at 300-500 pounds on the ram.

The weight of composition may vary between 5 gms and 25 gms per rocket depending upon the diameter. It is not possible to be more specific, since the ingredients vary in moisture, particle size and so on, from place to place and from different suppliers. Indeed most manufacturers devise their own formula to suit the particular type of case and charging pressures, and stick to this as closely as possible. In the end, firework rocketry is largely a matter of trial and error. Most signal rockets are of the end burning type, but the burning surface is frequently modified in shape to give better burning characteristics. As the charge burns away the chamber will, in effect become larger and this will cause a fall in gas pressure. In order to overcome this disadvantage, the end of the charge is shaped in such a manner that the burning surface will increase, and so tend to keep the chamber pressure constant. Cruciform or concentric channels are the two simplest and most favored designs for this purpose.

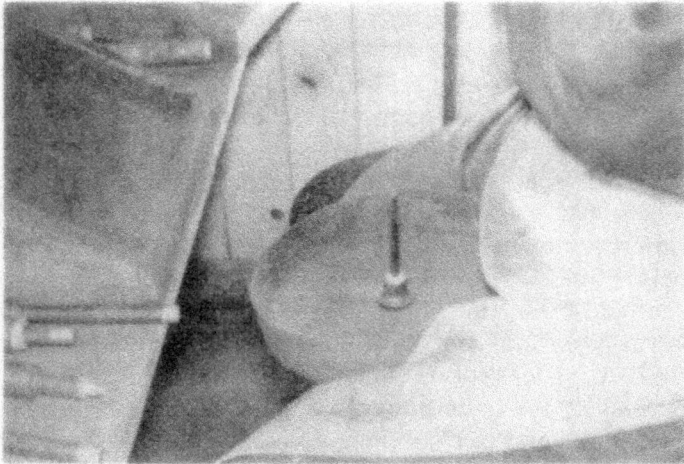

**Fig. 45a**   Rocket spindle. (Courtesy Brock's Fireworks.)

**Fig. 45b** Charging a rocket. (Courtesy Brock's Fireworks.)

**Fig. 46** Pressing in a mold. (courtesy Brock's Fireworks.)

The second type of rocket has a charge which is perforated along its long axis,. The perforation takes the form of a long thin cone, and this is usually formed during the charging process. The choked rocket case is placed over a tapering hard steel spindle which normally extends into the case for 3/4 of its length excluding the choke. Fig. 45a & 45b.

The diameter of the base of the spindle is about 1/12 of its length, and this should be the dimension of the rocket choke. The base of the spindle is supported by a solid concrete or steel post. For rockets of 1″ diameter and above, the composition is divided into 14 equal increments, each increment is poured separately into the case and consolidated by means of a hollow drift or ram which is subjected to a pressure of 2 tons. The case should be supported in a mold or tube to prevent collapse. (Fig. 46)

When the column of composition has reached the top of the spindle, the remaining increments are pressed with a stepped drift to form a plug with an axial hole. (Fig. 47) The finished rocket will have the appearance of Fig. 48 in cross section.

Smaller rockets of 3/4″ diameter and under, should have the composi-

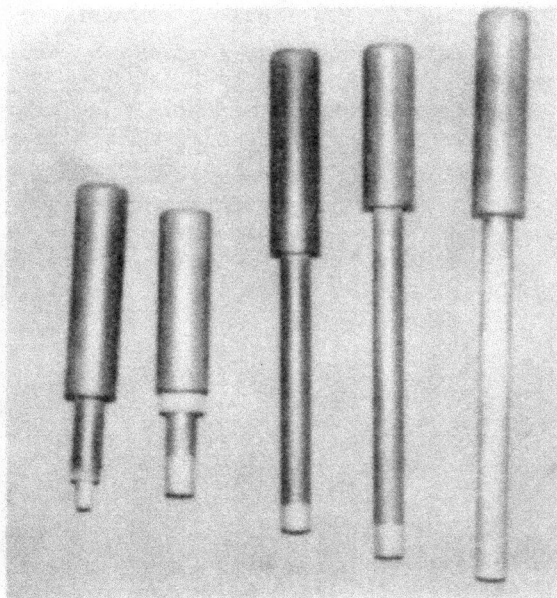

**Fig. 47** Rocket drifts. (Courtesy Brock's Fireworks.)

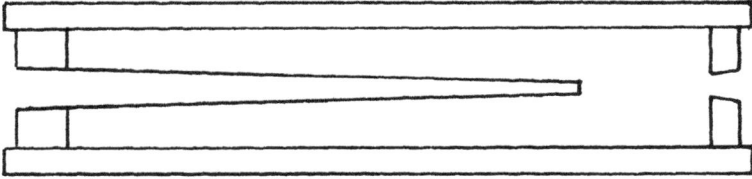

**Fig. 48**

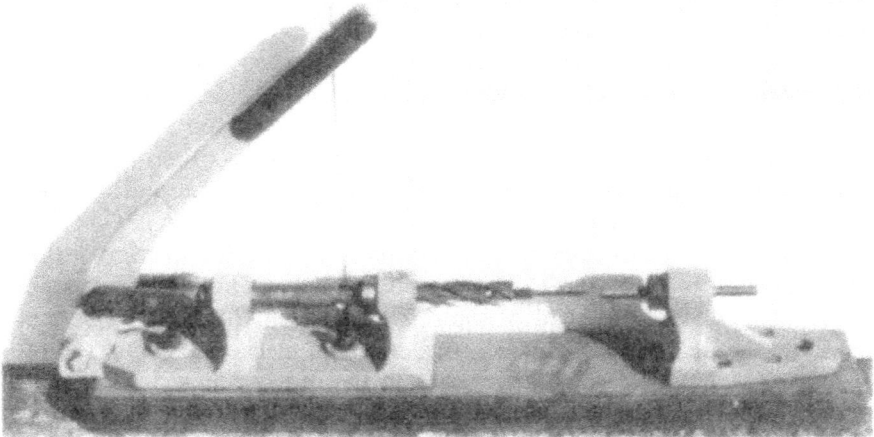

**Fig. 49** Rocket boring machine. (Courtesy Brock's Fireworks.)

tion charged in 8-10 increments. Very small rockets may be filled lightly with powder, and then bored by machine. This operation also serves to consolidate the composition. (Fig. 49)

Compositions for this type of rocket are of the sulphur, charcoal, potassium nitrate mixture type. The proportions vary widely, 60%-70% potassium nitrate, 15-20% charcoal, and 15 - 20% sulphur. It is not possible to be more specific in view of the problems mentioned above, but as a guide, rockets of 1″ diameter and over usually require a composition containing 63-67% potassium nitrate and this percentage increases as the rocket diameter decreases.

Rockets should be kept in a dry atmosphere, moisture absorption causes the composition to crack, and so gives rise to an increased burning surface, and explosion on ignition. Both types of rocket are ignited by means of a length of match inserted through the choke, or by means of fine grained gunpowder which is caused to adhere to the choke by means of a small quantity of thin shellac varnish. With both types of rocket, the flash of ignition should be sufficient to cause instant ignition of the burning surface, otherwise slow "take off" and erratic flight will result.

Stabilization of firework rockets is achieved by means of a stick. Many rules for stick dimensions have been given over the years, but the writer has found that it is only necessary to ensure that the centre of gravity lies a little to the rear of the head. The stick holds the rocket in a stable position during take off, and tends to counteract deviation during flight. Sticked rockets will always tend to turn into the wind due to unequal pressure on the stick.

Finned rockets are usually used for signal purposes. Here, the rocket is usually given sufficient forward velocity, to maintain stability during take off, by means of a small charge contained in the launcher. Fin design problems are covered fully in the technical literature. Line carrying rockets are stabilized by the harness and line and are also launched by a powder charge.

Firework rockets may carry a payload of stars, hummers or other pyrotechnic articles. The load is usually carried in a lightly rolled paper case attached to the head of the rocket. A flash from the propellant charge ignites a few grains of powder which ejects and lights the effects. The payload weight should be in the ratio.- 5 load to 3 propellant for a well constructed rocket.

# Chapter 11

# Drivers, Saxons, Tourbillions

*Drivers*

Wheels and moving fireworks are operated by rockets or more commonly by drivers. Drivers are stout tubes from 1/2″ to 1 1/2″ internal diameter, choked down to about 1/3 of a diameter and charged with a fierce composition containing a high percentage of gunpowder. The composition is charged solidly in small increments and does not contain the central spindle which is characteristic of rockets. The tubes are choked with a clay washer or by "pulling in" the tube in a choking machine.

Small wheels are turned quite adequately with tubes which are 1/2″ or 5/8″ in diameter and about 4″ long. A 5/8″ tube for example would be placed on a nipple with a central spiggot about 3/16″ in diameter. A small amount of clay is first charged around the spiggot to produce a washer, and then increments of composition are added and charged with several light blows. With fierce compositions, many light blows are better than a few heavy ones, and increments of composition should be sufficient to occupy one diameter in height. A charge of clay is finally rammed into the tube to close it up, though not all tubes are clayed, for some are just fitted with a paper cap to enable fire to be transferred to a new driver where relays are required.

The following compositions are suitable for small drivers:-

|                      | A  | B  | C  | D  |
|----------------------|----|----|----|----|
| Meal gunpowder       | 68 | 80 | 88 | 70 |
| Charcoal 150 mesh     | —  | —  | 12 | —  |
| Charcoal 40/100 mesh  | 7  | —  | —  | —  |
| Potassium nitrate     | 15 | —  | —  | —  |
| Aluminum "Bright"     | —  | 20 | —  | —  |
| Iron Turnings 60 mesh | —  | —  | —  | 30 |
| Sulphur               | 10 | —  | —  | —  |

The smaller drivers are used for a variety of different types of wheels

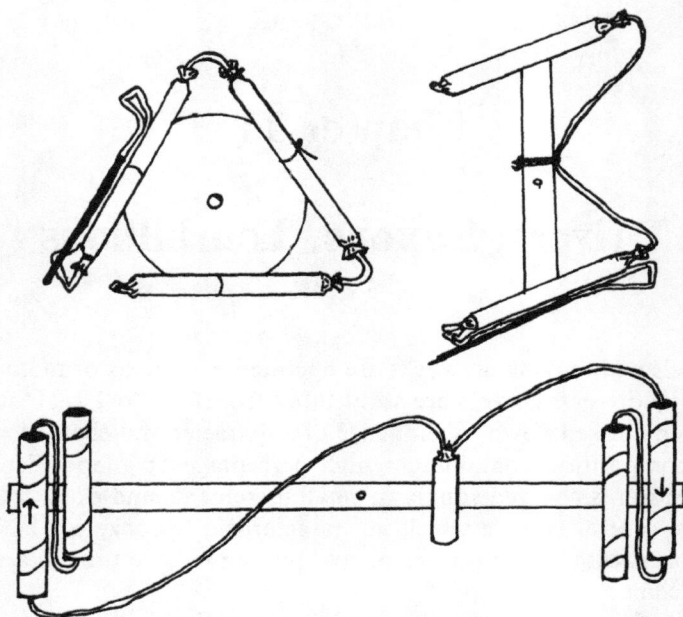

**Fig. 50** Small wheels

as the diagrams in Fig. 50 show. The older wheels were made with a central boss, three or four spokes and a beech hoop, but these have been replaced in more recent times with constructions made out of plywood and hardboard.

Large drivers are used for the bigger display wheels from six to fifteen feet in diameter. The tubes are 1″ or 1 1/2″ in diameter and about 9″ long. Compositions are a shade slower and often contain more charcoal than the slower sizes e.g.

|                          | A  | B  | C  |
|--------------------------|----|----|----|
| Meal gunpowder           | 74 | 65 | 65 |
| Potassium nitrate        | 10 | 5  | 5  |
| Charcoal 40/100 mesh     | 6  | —  | —  |
| Charcoal 30/60 mesh      | 6  | —  | —  |
| Sulphur                  | 4  | 5  | 5  |
| Iron Turnings 20 mesh    | —  | 25 | —  |
| Titanium Turnings 40 mesh| —  | —  | 25 |

*Saxons*

These fireworks are used in combination with a small color case to produce small colored wheels which are usually part of a larger design.

Thick walled tubes are charged with composition, but the tube is closed with clay at both ends, and the fire issues from a hole which is cut into the side of the tube at right angles to the axis of the tube. The hole is near to the clay plug which closes one end, but the size and position of this hole depends on the strength of the composition. The hole is often bored 1/2″ or so from the clay plug so that the composition burns all around the hole when the tube is ignited, thus giving an extra large thrust at the beginning to get the wheel started. Unfortunately it is a common sight to see sluggish saxons which do not seem to have enough power to get started, or else they become stuck with a piece of match piping wrapped around them or have a nail which is too tight. Another snag with saxons is the fact that the hole in the tube gets larger during the burning process with the inevitable slowing of the piece. Thick walled tubes about 5/8″ in diameter with a 3/16″ wall or 1/2″ in diameter with a 3/16″ wall are used, and they are about ten to twelve inches long.

Compositions should not be too hot, i.e. they should not contain too much potassium nitrate and sulphur or else they are liable to burn the jet hole away too much with the resulting loss of force. Old mixes

Fig. 51   Saxons.

**Fig. 52**   Double ended saxon

were rather hot burning e.g.:

| Meal gunpowder | 50 |
| Potassium nitrate | 25 |
| Sulphur | 25 |

but the following are more typical and useful:

| Meal gunpowder | 50 |
| Potassium nitrate | 30 |
| Charcoal 40/100 mesh | 10 |
| Sulphur | 10 |

a little titanium can be added to this mixture of course. Saxons are made in various forms, but are mostly double. A tube about 12″ long has 1/2″ of clay solidy rammed into one end and then is charged with 5″ of composition. 1″ of clay is then charged, followed by 5″ of composition and a final 1/2″ of clay. The tube is now full. A hole to receive the nail is bored through the middle of the tube, through the clay, and then the jet holes are bored near the ends, but on opposite sides of the tube as in Fig. 52.

Saxons can be fired simultaneously from both ends or they can be arranged to fire successively by inserting a piece of quickmatch in the base of the composition of one end and so transferring the fire to the opposite end as in the diagram above.

Some manufacturers make the saxon in two separate units, joining the two together in the center with a paper tube or a piece of wooden dowel. Single saxons are also made as in the diagram below. (Fig. 53).

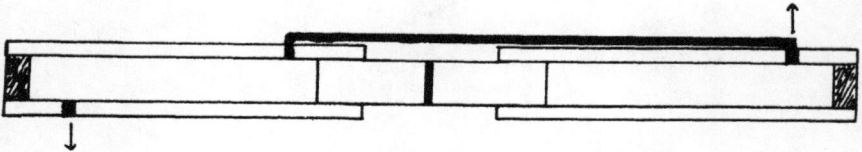

**Fig. 53**   Two single saxons attached to a piece of wooden dowel.

A small spacer tube should be fixed between a saxon and the timber-work to prevent the saxon from striking the woodwork during the performance. Color cases are usually fixed on the side of the saxon to produce the fire rosette. They are timed to burn the same length of time as the saxon and require careful fixing with glue and paper or else they are liable to fly off during the performance.

*Tourbillions*

The French word "tourbillion" or "whirl wind" is still used to describe these fireworks which have appeared in many forms and sizes. The older and larger forms were in effect a fierce type of saxon which first of all spun around on the ground and then lifted up into the air by means of two or four holes bored in the underside. (Fig. 54).

A curved stick roughly the same length as the tourbillion itself was nailed to the centre of the tube, but at right angles to the tube. The process of boring the six holes in a large tourbillion is quite a tricky one and needs to be carefully done. The holes in the side and base also need to be exactly at 90° to each other or else the performance will be erratic. A useful way of making the tube is to slip it inside a hollow metal or paper tube which already has the holes drilled in the correct positions, thus using this external tube to mark the position of the holes prior to drilling. It is also unwise to drill tubes which already contain composition a much better method being to drill or punch the

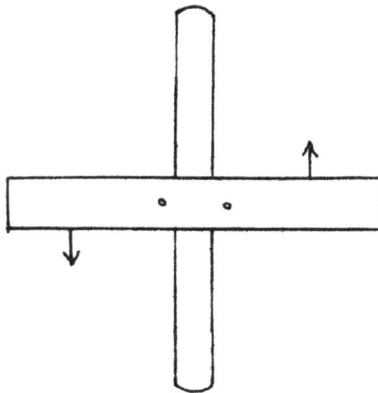

**Fig. 54** Position of jet holes in the side and base

holes in the empty tube and cover them with a piece of paper or adhesive tape which can be removed after filling. This also applies to
saxons of course.

The larger sizes are made with a tube which is approximately 1″
diameter, 8″ long and with a wall thickness of 9/32″. The holes are
about 3/16″ depending on the strength of the composition, but they
also must be carefully spaced so that there is roughly the same quantity
of composition on either side of each hole. The following composition
is a good one though some adjustment may need to be made with the
diameter of the jet hole.

| Potassium nitrate | 45 |
|---|---|
| Meal gunpowder | 35 |
| Sulphur | 5 |
| Charcoal 40/100 mesh | 15 |

Sticks are usually made out of a piece of beech hoop of the type used
for sieves and large wheels, and are fixed to the tube with a copper
nail. The tourbillion is matched in such a way that the match is ignited
on the top in the center, transferring the fire first to the two side jets
for spinning and lastly to the lifting fire jets at the base. The match is
fixed with tape and then covered with paper. Small tourbillions are
made with a tube about 1/2″ in diameter, about half full of composition. They are made to ascend with a single jet punched at about 30°
or 45° which both lifts the firework and spins it at the same time (Fig.
55). These small articles use much fiercer compositions with a larger
percentage of mealpowder and as before, the size of the jet hole is related to the burning speed of the composition. Wings are usually made
of paper board or plastic. Closely related to this type of firework is the
famous Flying Saucer. This piece has various names such as Geyser,
Girandola, Flying Crown and consists of circular horizontal wheel
which spins at great speed on a spike at the top of a pole and then
takes off into the air to climb several hundred feet. Some varieties ascend and descend and finally re-ascend during flight.

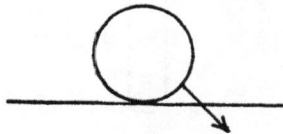

Fig. 55  Position of the jet hole

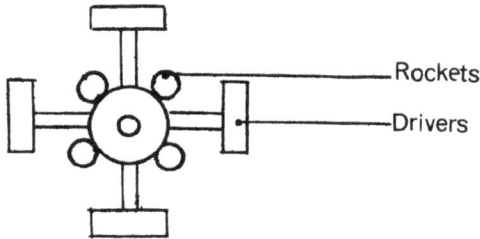

**Fig. 56** Flying pigeon

This piece is very difficult to produce, but it basically consists of very fierce horizontal drivers which cause the unit to spin on the post and a few fierce vertical drivers or rockets which lift the piece in the air.

No attempt will be made to describe the manufacture of these items for there are many pitfalls which only considerable experience can overcome. Those who attempt to make the pieces will find that either the drivers are not fierce enough to lift the piece or they explode. In addition badly made drivers will burn incorrectly, sending the saucer into the air at the wrong angle and possibly falling into the crowd. The saucer usually carries a small shell which fires at the highest point of the flight. It can also be fitted with fountains or small pot à feu.

*Flying Pigeon*

The line rocket or flying pigeon is a very popular item at firework displays and frequently consists of a wooden block with a hole through the center so that it can slide along a tightly stretched rope or wire. Four large rockets are fixed to the block and, in addition, there are four fierce drivers attached to a small hoop fixed on the block. See Fig. 56.

The first driver rapidly spins the unit on the rope, and as soon as it burns out it ignites the first rocket which transfers the piece to the other end of the rope. The burnt out rocket then ignites the next driver and the process continues, backwards and forwards along the rope. Where ropes are used, the friction is quite high and large rockets of at least 3/4″ bore need to be used. Whistles and colors are often added to the pigeon. A tube needs to be placed on each end of the rope to prevent the pigeon striking the post and thus being damaged. A good way of assembling the pigeon is to attach the rope to two firm stakes and then stretch it with two stretchers according to the diagram. The

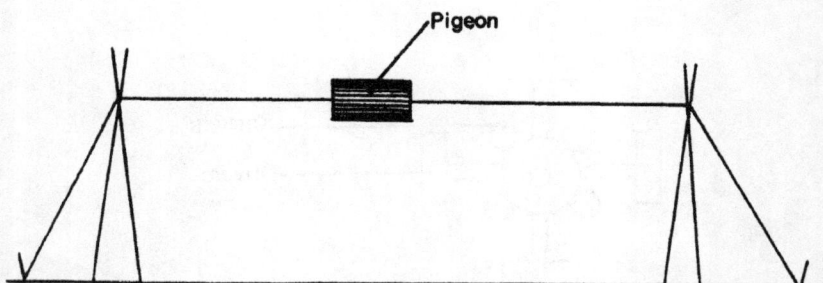

**Fig. 57**   Flying pigeon

stretcher consists of two wooden poles loosely joined at the top with a nut and bolt. (Fig. 57).

Smaller pigeon type fireworks called rats consist of a relay of very fierce drivers which follow each other down the wire. They are charged with a mixture of gunpowder and iron and the tubes are about 1/2″ in diameter.

# Chapter 12

# Shells

The greatest skill of the firework maker can be exhibited in the manufacture of star shells, or "Bombs" as they are frequently called. Shells are produced in huge quantities every year but the quality varies from, at one extreme, the most ordinary types, which literally drop the stars into the sky in a disorderly heap, to the magnificent pattern shells at the other extreme.

The older method of shell construction involved the manufacture of hollow paper spheres about 3/16″ thick but this was a costly and time-consuming business. Strips of paper were soaked in paste and then laid in layers in a hemi-spherical mold. A hemi-spherical former is used to press the paper into the mold in order to press the strips of paper down. When the shell halves are dry, the rough edges are trimmed with some type of lathe. (Fig. 58).

When the two perfect hemi-spheres are ready, a hole is punched in one of them to receive the fuse and the two halves are glued together. The final process consists of pasting strips of paper, which are well soaked in paste, all over the outside. Long strips are used, usually stretching from the fuse hole to the base.

The method outlined above is hardly ever employed to-day and the paper sphere is made up from four hemispheres, two inner ones and two outer ones. The hemispheres are pressed out of sheets of pasted strawboard cut out in the following shape. (Fig. 59).

Ideally, the thickness of the two inside shells should be equal to those

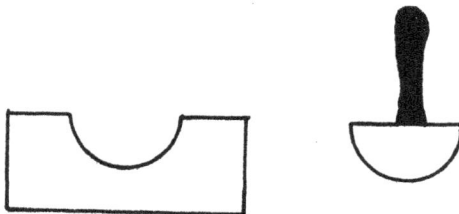

Fig. 58 Shell mold and former

143

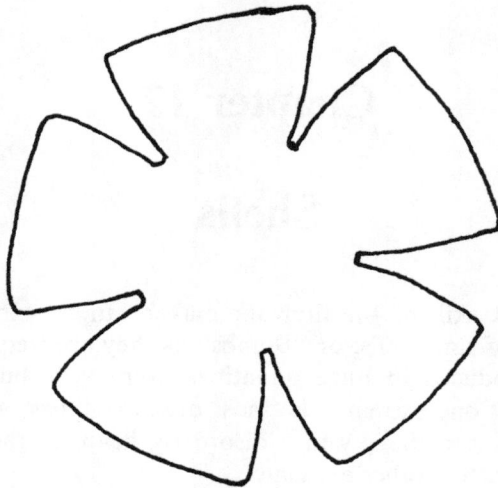

**Fig. 59**   Paper cut out for shells

on the outside, so that when the halves are in different planes, the shell should break at least into four quarters to produce an even burst.

In practice this even burst is not easy to obtain and instead of all the stars being thrown out evenly from the center, they tend to fall downwards in a kind of bouquet. The old method was even worse, for in effect the shell broke into two halves and dropped the stars out in a heap.

It will be seen that the formation of the shell wall is quite vital and that the pattern of the burst is entirely dependent on this in relation to the nature of the bursting charge.

The easiest type of shell to make is similar to the Japanese "Poka" page 266. The sphere is loaded with stars and 1/2 to 1 oz of meal gunpowder is poured loosely amongst them before the fuse is glued in. In some cases a mixture of gunpowder (85%) and fine charcoal (15%) is used instead, employing about twice the quantity one would use in the place of neat gunpowder.

Two methods are used to burst round shells. One method has been outlined already and consists of bursting a shell with a comparatively weak outside wall. The other method is to use the Japanese "Wari-mono" page 257 technique which requires the formation of an extremely hard exterior wall consisting of many layers of pasted kraft paper. This is adequately covered by Dr. Shimizu.

The shell maker thus has the choice of three methods of bursting round shells:-

1. Employ a small gunpowder charge and a weak exterior wall to obtain the bouquet effect.
2. Employ a much fiercer bursting charge and a comparatively weak shell wall in order to obtain an even burst. The use of a fierce bursting charge involves the risk of destroying the stars or causing ignition failures. Flash powder is sometimes used for this purpose, but there are many drawbacks and dangers to the technique.
3. Employ a very strong outside shell wall and use a comparatively slow expanding bursting charge. In the west, various grades of grain powder are used for this purpose but in the east they have their own distinctive techniques outlined in chapter 19.

Round shells are used in various sizes, the smallest being made to fit a 2″ diameter tube and the largest 36″. The most common sizes are 3″, 4″, 5″, 6″ and 8″, these being fairly universal in spite of the fact that they are invariably measured in millimeters. 12″ and 16″ shells are sometimes manufactured but it is a by-word among firework makers that their effect is frequently no better than a 6″ or 8″ shell. Dr. Shimizu reports that two 36″ mortars are known to exist in Japan. It appears that this unusually large size is not regularly manufactured but has been a feature of a yearly summer display at the city of Nagaoka. It is of interest to note that it has been said of these shells also that they are not as attractive as one might think.

In the last few years plastic ball shells have been introduced, the molded plastic being made in such a way that the delay and lifting charge are all built into the same unit. They are cheaper and easier to fill, though in some respects paper still seems to be superior.

Some manufacturers describe their shells in terms of circumference instead of diameter. Fortunately the practice is not common and is possibly a device to deceive the uninitiated. 12″ sounds much better than 4″, but what a disappointment when you receive the shell.

After the shell has been filled with stars and blowing, the fuse is carefully glued into place, but care must be taken to see that the fuse is secure or else it may be forced through into the shell by the lifting charge. Lifting charges vary, depending on the size of the shell, the quality of the powder, the length of the mortar and the tightness of the fit of the shell. A 4″ round shell would take about 2 oz of the British T. P. Cannon powder while an 8″ shell would require about 9 oz. The

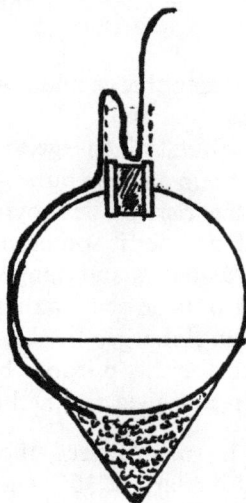

**Fig. 60**   Cross-section of a round shell

lifting charge is usually placed in a cone though any substantial paper
container would be adequate. The cone is particularly useful in situations where mortars are to be re-loaded, for the cones fit into each
other and do not take up too much room in the tube. (Fig. 60).
Lastly a length of piped match is tied into the fuse and continued down
to the lifting charge so that the two will be ignited almost instantaneously. It is advisable to have two pieces of match between the time fuse
and the lifting charge, for a failure in this position could cause the
shell to explode in the mortar. The shell fuse itself is usually wrapped
with about three turns of cotton cambric to hold the match pipe, for
shells are often carried by their leaders and cambric will not tear as
paper does. The black match at the end of the match pipe is usually
covered with an empty lance tube for safety.

Three types of shell fuse are made. The older type consisted of a very
hard tube about 3/4″ in diameter, about 1 1/8″ outside diameter and
about 1 1/2″ long. A white fire composition such as the following:-

| | |
|---|---|
| Mealed gunpowder | 65 |
| Potassium nitrate | 25 |
| Sulphur | 10 |

was charged into the tube in about eight increments of 1/8″ when charged, each increment being rammed with many light blows until it is rock hard. 1″ of composition burns about 4 seconds. Solid filling is essential or else there is a good chance of getting a "blow through". This fuse burns with a white light and is quite effective.

The second method uses tubes about 3/8″ or 5/16″ in diameter and about 2 1/2″ long. Meal gunpowder, or better still a soft unpolished grain powder, is charged into the tube in the manner described above until there is about 1″ or 1 1/4″ of solid powder. The remainder of the tube is filled with lengths of quickmatch which project about 1/2″ outside the tube. The match is held in place through wedging it by breaking down the inside wall of the tube with a piercer, or tying it into a piece of cambric which is glued on to the end of the tube.

A third way of making fuses is simply to employ a piece of Bickford or other safety fuse. These fuses burn at the rate of about 3 seconds per inch and vary in thickness from 1/4″ to 3/8″. In England it is usual to glue two 1″ lengths of Bickford into a small wooden plug. The Americans tend to use one piece of 3/8″ Bickford fastened to the center of a mill board disc. The ends of the Bickford require to be primed with gunpowder paste, or better still pierced and threaded with a short length of quickmatch, to ensure ignition. (See Fig. 61).

The firing tubes or mortars for the shells consist of strong steel or paper tubes. The length of the tube varies with the type of shell, but mortars are normally 12″ to 20″ for round shells or short cylinders, and 30″ to 40″ for long cylinder shells. Paper tubes are adequate up to about 5″ diameter and should be stout, straight rolled tubes about 1/2″ thick, fitted with a strong wooden base firmly nailed in position. Spirally wound tubes can be used but are less satisfactory. Cylindrical shells should always be fired from steel tubes but maroons or salutes should be fired from paper tubes in case they explode in the tube. Steel

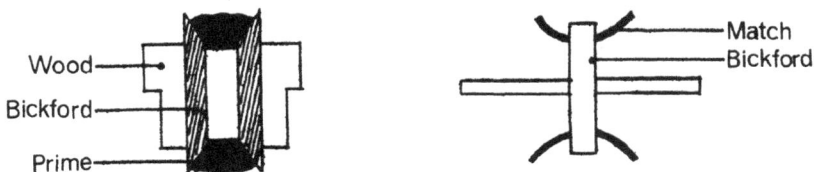

**Fig. 61** Left: Bickford in wooden fuse. Right: Bickford in cardboard disc

mortars are made out of ordinary steel tube fitted with a firm steel base which is carefully welded in position. Cast iron should not be used of course.

Mortars for small shells are frequently placed in long wooden boxes which must be very securely held with stakes to prevent them from falling over during the firing operation. Larger tubes are buried in the ground or secured in drums of sand. A thick wooden board should be placed under mortars, for they tend to dig themselves in the ground and can be very difficult to remove from the wet earth or clay.

Sensitive compositions should have been eliminated by now, but where they are used, it is a wise precaution to place sandbags between the mortars and the operator or crowd for detonations have caused serious accidents. Flash charges containing chlorates and sulphur or sulphides in large cylinder shells are particularly liable to detonate and (in the opinion of the writer) should be illegal in all countries.

Izzo(8) and Davis(7) have described the manufacture of cylinder shells and it must be said that some of the best effects are produced by them. Unfortunately they are also the most expensive items because of the tremendous amount of hand work which goes into the more effective ones.

To make a cylinder shell, a former and some cardboard discs are required, the diameter being about 1/2" less than that of the mortar. A sheet of medium unglazed kraft is then cut so that it is about 2 1/2" or 3" longer than the length of the completed shell and sufficiently long to go about three times around the former. Sometimes, a thicker sheet of paper the length of the finished shell is wrapped inside also. The paper is rolled around the former and pasted just on the outer edge to form a bag. A disc is placed in the bottom of the bag and the bottom 1 1/2" of paper is cut into 3/4" strips which are then turned over as in (Fig. 62).

The bag is then filled to within 1 1/2" of the top with stars or other

Fig. 62  Shell bag

**Fig. 63** Filling the stars and powder

effects, carefully consolidating the stars but not breaking them. Gunpowder is next poured into the stars to fill up the spaces so that the bag is quite full. This is a necessary operation since the shell wall is flimsy and gains its strength only from the filled interior. In special shells, the powder is only placed in the center. In this case the stars are filled around a hollow brass or copper tube in the center. Powder is then placed in the center of the tube which is slowly withdrawn, leaving the powder in the middle. (Fig. 63)

The next operation is to add the fuse which has been previously glued into the center of a disc. This disc is placed in the bag on top of the stars and gently tapped down. The shell wall is then cut into strips as at the beginning, turned down and held in position with another disc. (Fig. 64)

Stringing takes place next. The thin string is soaked in paste and wound around the shell first down the length of the shell and secondly around the sides so that the whole shell is covered in squares with a side of anything from 1/4″ to 1″ depending on the type of shell. The stringing must be even and quite taut to produce an even burst. (Fig. 65).

The final operation consists of covering the outside of the shell with two or three layers of pasted kraft paper. Small shells require only a small quantity of paper but larger ones take more. The shells are then completed by adding the lifting charge and leaders.

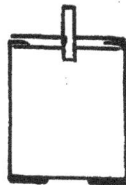

**Fig. 64** Filled bag

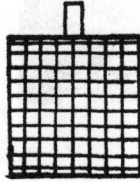

**Fig. 65**   Shell after stringing

Cylinder shells are the best ones for carrying smaller tubes such as whistles, hummers, tourbillions, and flash charges, for they are more economical of space. It is usual practice to place the special effects in the bottom of the shell with a charge of stars on top. Shell tourbillions are usually about 1/2″ in diameter, 4″ long and drilled with two holes, but of course they require no stick. Flash charges are usually thick tubes with wooden end plugs, one plug being furnished with a length of Bickford fuse. Sometimes a number of small shells are placed inside a large shell, so that after the initial burst of stars, the smaller ones

**Fig. 66**   Shell manufacture. (Courtesy Pains-Wessex, Ltd.)

**Fig. 67** Round shells. (3-12 inches.)

burst in a ring below. Shells are also charged sometimes with short, quick firing roman candles.

Cheap cylinder shells are sometimes manufactured out of cannisters fitted with cap ends and two discs. (Fig. 69).

The effect is not so good as the more complex cylinder shells, but is

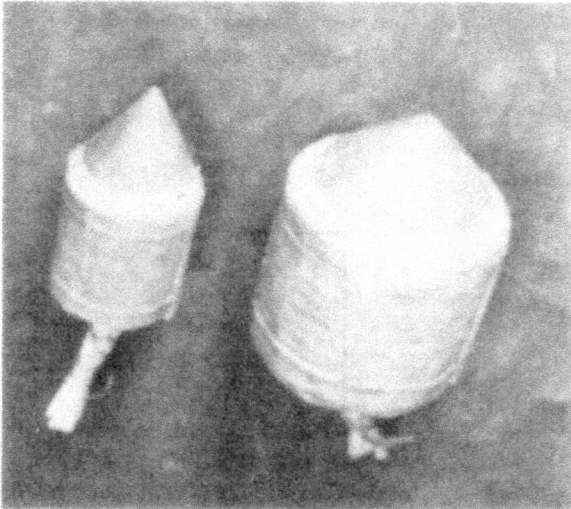

**Fig. 68** Cylinder shells. (5 and 8 inches.)

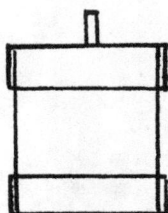

**Fig. 69** Cylinder shell with cap ends.

quite useful for ordinary work. The older methods employing tough
cylinders with wooden ends are not commonly used now; they are too
complex and expensive.

Repeating shells are somewhat more complicated but are made
basically in the same manner as ordinary cylinder shells though there
are two techniques. One technique involves making short cylinders
all of the same diameter and height and marrying them to each other
so that the fuse of one projects into the section above. The sections are
all strung together, this being a vital operation for success.

The sections of repeaters are usually fitted with two second fuses.
The first fuse may also be of two seconds if the shells are to burst on
the way up; otherwise it is four of five seconds where the shell reaches
its fullest height before the initial burst. Needless to say the lifting
charge increases with the number of repetitions and shells bursting on
the way up require larger lifting charges than the others. (Fig. 70).
The second technique for repeating shells is the one more commonly
used in Europe and involves the placing of one bomb inside the other.

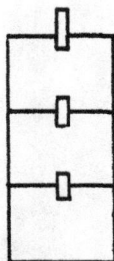

**Fig. 70** Repeating shell

Fig. 71   Repeating shell

The last section to burst is frequently a flash maroon which is carefully made and strung prior to being placed in the bottom of the bag of the second section. Section two when complete then goes into the bottom of the bag of the first burst. (Fig. 71).

Careful packing and stringing is essential or else the inside shell is destroyed by the main shell burst. It also means the use of many different sizes of discs and formers. Fig. 72 shows a set of repeating shells.

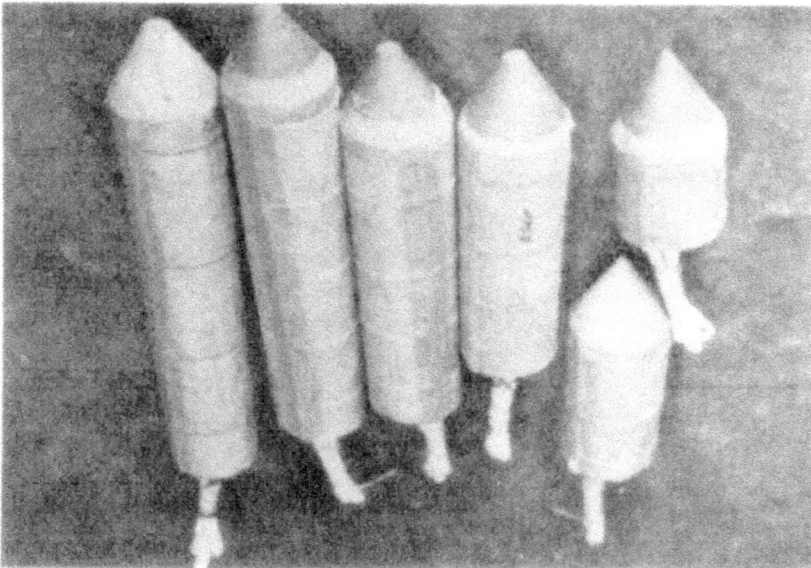

Fig. 72   Set of repeating shells

Further detail on shell manufacture is pointless because every shell maker has to adjust his technique to fit the materials in hand and the unreliability of fireworks makes him adhere to those methods he has tested in his own experience. Again, a list of types of shells is pointless since the numbers of permutations are limited and obvious.

Perhaps the 36″ Japanese shell is the most unusual of shells and we are grateful to Dr. Shimizu for this photograph of such a shell being loaded for the Nagaoka festival by the Kashiwazaki Fireworks Co. (Fig. 73).

The comet shell is an interesting variation. A large pressed comet is fastened on to the end of the shell so that it will produce a spiral of fire as it rises in the air. The Japanese glue a small comet on to the out-

Fig. 73   Loading a 36″ shell

Fig. 74   Small comet bomb

side of some of their shells, but they also sometimes coat the whole of the outside of a shell with a streamer composition.

A smaller version of the comet bomb is to press a star into the bottom of a tube about 1 1/4″ in diameter and then place stars on the top of it. (Fig. 74).
Small units like these are also fired out of roman candles.

Parachute shells are made in much the same manner as other shells, though the bursting charge has to be cut down and the parachute needs to be protected with sawdust. A good way is to tie the flare into the small bursting charge which is in turn tied on to the inside of the fuse. The rest of the shell is filled with the parachute and sawdust. (Fig. 75).

Parachutes for ordinary work are normally made of cotton cambric or good quality tissue paper. 15″ of twine is tied on to each corner, and these are tied together at their ends and fixed to about 12″ of wire or asbestos string. This last 12″ needs to be wire or asbestos so that it will not be burnt by the flare. When it is required to make a string of

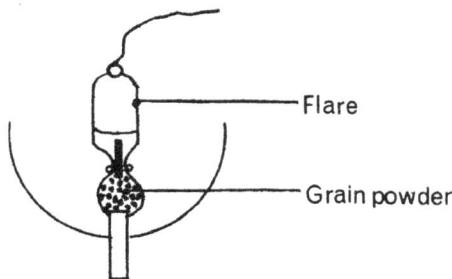

Fig. 75   Parachute shell

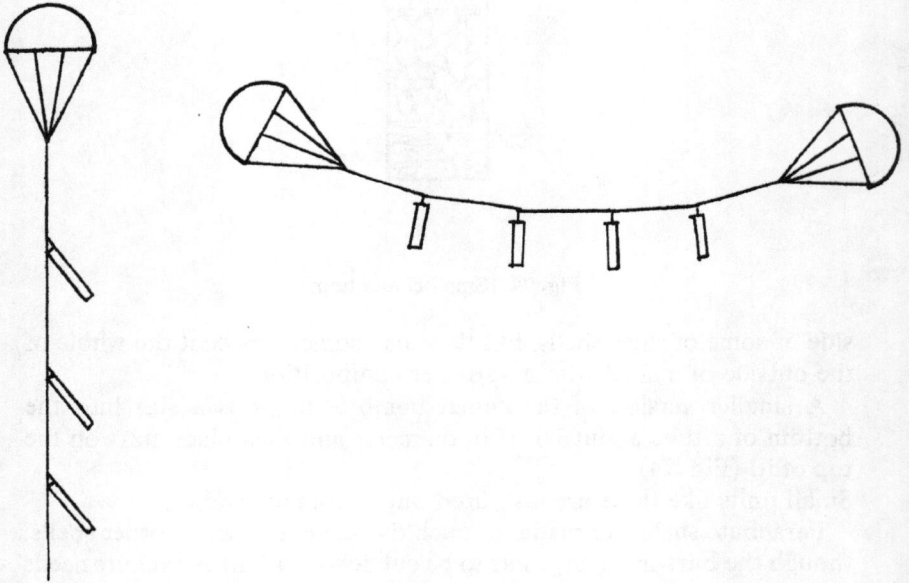

Fig. 76   Parachute chains or festoons

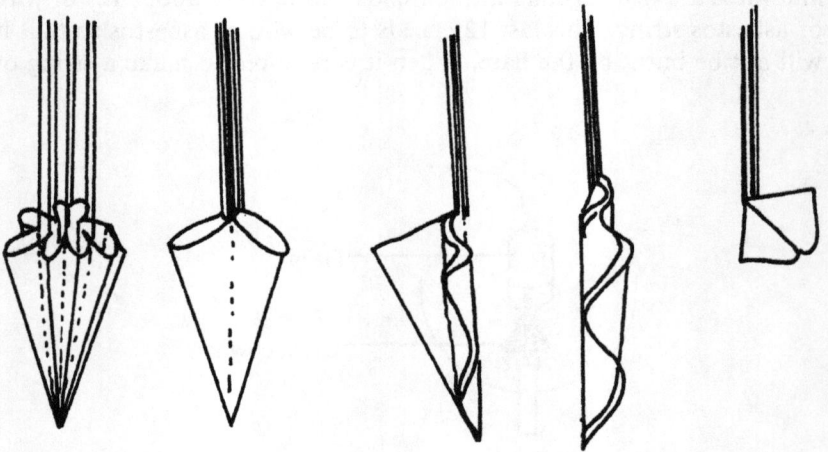

Fig. 77   Method of folding parachutes

lights, there must be as much as three feet of string between each light; otherwise they appear to be close together in the sky. Sometimes festoons are made by attaching parachutes to both ends of a string of lights. (Fig. 76)

Parachutes must be specially folded so that they open correctly. The following diagrams illustrate the procedure and it is wise to sprinkle the chute with talcum powder before folding it up.

The folding procedure applies equally well for rockets and shells. (Fig. 77).

# Chapter 13

# Gerbs, Fountains, Rains, Squibs, Cones

*Gerbs*

The French word "gerbe" meaning a spray or sheaf of corn is still used for firework fountains, particularly those which produce force in choked tubes. A wide variety of fountains can be found within the firework range but generally speaking there are two basic types; those of the long narrow variety and the short squat fountains. The mixtures naturally vary immensely with the type of tube and method of filling. The long narrow "Fix" used for "feu croisée" in set pieces for instance, requires fierce compositions which need to be carefully charged, while at the other extreme, some short squat fountains are so slow burning that the composition requires little more than being poured into the tube and shaken down.

*Fountains*

Tubes for fountains are usually about 9″ long and from 1/2″ to 1″ in diameter. There are also other sizes which are smaller or larger than the ones mentioned. The wall thickness is sufficient to prevent the fire bursting the side of the tube. Long narrow fountains are either choked with a clay washer to one half or a third of a diameter or just "pulled in" as described in chapter 5. Pulled in tubes are sometimes essential; some glitter fountains for example will not function correctly in clayed tubes, mainly because the solid burning products do not seem to be able to get past a clay choke, with the consequent "coking up" of the tube.

The so-called display "Fix" are 1/2″ or 5/8″ inside diameter and 9″ long. They are choked to half a diameter and charged by hand or pressed in about eight increments. The compositions are quite fierce for they are required to produce two or three feet of concentrated fire for set piece pattern work. The following compositions are used for this purpose:-

|                     | A  | B  | C  |
|---------------------|----|----|----|
| Meal gunpowder      | 84 | 73 | 64 |
| Charcoal 150 mesh   | 16 | —  | —  |
| Potassium nitrate   | —  | —  | 8  |
| Sulphur             | —  | —  | 8  |
| Iron 60 mesh        | —  | 27 | 20 |

Fountains of 3/4″ to 2″ inside diameter require less fierce mixtures but if they are not strong enough they will "go off" lower down the tube owing to blockage from dross. These compositions are usually cheaper because they contain less gunpowder, but they are invariably dirty because finely powdered materials are required. The following mixtures are fairly uncomplicated

|                              | A  | B  | C  | D  |
|------------------------------|----|----|----|----|
| Potassium nitrate 150 mesh   | 72 | 40 | 44 | 44 |
| Sulphur                      | 8  | 8  | 9  | 9  |
| Charcoal 40/100 mesh         | 16 | —  | —  | —  |
| Charcoal 28 mesh             | 4  | —  | —  | —  |
| Charcoal 150 mesh            | —  | 8  | 8  | 8  |
| Meal gunpowder               | —  | 16 | 9  | 9  |
| Coated iron 20 mesh          | —  | 24 | 30 | —  |
| Titanium 40 mesh             | —  | —  | —  | 30 |
| Aluminum flitter 10/30 mesh  | —  | 4  | —  | —  |

The old English "Flower Pot" is produced from a combination of red orpiment and lampblack. The burning is very drossy with the result that bits of dross are forced into the cool air where they sparkle with their characteristic golden coruscations. The following composition is typical, but it needs to be emphasized that only certain types of gas black will function correctly and the material has to be chosen by trial and error.

| Potassium nitrate 150 mesh | 53 |
|----------------------------|----|
| Sulphur                    | 23 |
| Red orpiment               | 7  |
| Meal gunpowder             | 7  |
| Lampblack                  | 10 |

Sometimes this particular composition functions better when it is charged with a funnel and wire than when it is charged with a mallet.

Silver fountains are made with titanium or aluminum. Titanium is uncomplicated, but aluminum powder varies tremendously from one manufacturer to another. Fountains made with aluminum often start off well and then get blocked up with molten dross lower down the tube. As a general rule it is necessary to use two or more grades of aluminum in the one composition and keep the quantity as low as possible.

| | |
|---|---|
| Meal gunpowder | 72 |
| Potassium nitrate 150 mesh | 7 |
| Charcoal 40/100 mesh | 7 |
| Aluminum dark pyro | 7 |
| Aluminum 80/120 mesh | 7 |

A rather more difficult fountain is a silver one using barium nitrate as the oxidizer. The tube is about 3/4" in diameter and 6" long. The choke must be of clay (about half a diameter) and the wall of the tube must be very strong and possibly lined with asbestos, for the temperature is very high indeed. Mixtures of barium nitrate and aluminum are of course sensitive and so they should not be charged with a mallet and drift, but pressed. Again, everything depends on the aluminum for success. The following mixture represents a rough guide only

| | |
|---|---|
| Barium nitrate | 45 |
| Potassium nitrate | 5 |
| Meal gunpowder | 5 |
| Aluminum | 45 |

The aluminum here is a mixture of dark pyro, bright polished and flitter grades.
The glitter or flitter gerb is just a modified glitter star composition e.g.

| | |
|---|---|
| Meal gunpowder | 68 |
| Antimony sulphide | 14 |
| Sodium oxalate | 11 |
| Aluminum "Bright" | 7 |

The yellow glitter star composition works quite well in a straight tube without choke.

Short squat fountains are 1/2" to 1" in diameter and 2" to 6" long. They are frequently not choked, but in some case they may be fitted with a tight fitting cardboard washer instead of a choke. Mixtures vary enormously with the method of filling, but most of them require some degree of consolidation, perhaps with light hand pressure, of funnel and wire type filling.

**Fig. 78**   Gold or silver rain

Compositions for unchoked tubes:-

|                              | A  | B  | C  |
|------------------------------|----|----|----|
| Meal gunpowder               | 56 | 45 | 66 |
| Potassium nitrate 150 mesh   | 22 | 22 | —  |
| Charcoal 40/100 mesh         | 11 | —  | 20 |
| Aluminum dark pyro           | 6  | 11 | —  |
| Aluminum 80/120 mesh         | 5  | —  | —  |
| Sulphur                      | —  | 11 | —  |
| Titanium 20/40 mesh          | —  | 11 | —  |
| Charcoal 150 mesh            | —  | —  | 7  |
| Iron coated 60 mesh          | —  | —  | 7  |

Compositions for choked tubes:-

|                              | A  | B  | C  |
|------------------------------|----|----|----|
| Meal gunpowder               | 42 | 17 | 15 |
| Potassium nitrate 150 mesh   | —  | 46 | 45 |
| Charcoal 40/100 mesh         | —  | 17 | —  |
| Sulphur                      | 7  | 3  | 6  |
| Charcoal 150 mesh            | 10 | —  | 9  |
| Iron coated 60 mesh          | —  | —  | 8  |
| Antimony sulphide            | 20 | —  | —  |
| Sodium oxalate               | 14 | —  | —  |
| Aluminum "Bright"            | 7  | —  | —  |
| Charcoal 28 mesh             | —  | 17 | —  |
| Lampblack                    | —  | —  | 12 |
| Aluminum flitter 10/30 mesh  | —  | —  | 5  |

*Rains*

Gold and silver rains consist of long narrow fountains of small bore which are fierce enough to produce a good show of sparks. The tubes are about 1/4″ or 3/8″ in diameter and from 3″ to 5″ long. As they are cheap items they are invariably dry rolled, being merely pasted at the edge of the paper. Tubes are charged with a funnel and wire or other means, but it should be pointed out that dry compositions containing more than a few percent of aluminum should not be used in a funnel and wire apparatus. Fig. 78 shows a drawing of a Rain. The following compositions are useful:-

|                       | A  | B  | C  |
|-----------------------|----|----|----|
| Meal gunpowder        | 75 | 75 | 80 |
| Charcoal 40/100 mesh  | 25 | 23 | 5  |
| Aluminum dark pyro    | —  | —  | 5  |
| Aluminum "Bright"     | —  | 1  | 5  |
| Aluminum 80/120 mesh  | —  | 1  | 5  |

Short rains for rockets and shells are occasionally charged as short lances.

Squibs are no longer manufactured in Britain, but they were in effect golden rains with a small gunpowder "bounce" at the end. They also had a practical use, for they were sold to householders who placed them inside coal-burning oven "flues" to discharge collections of soot from places difficult to reach by other means.

*Flying Squibs*

In effect these were fierce little drivers, but they are no longer sold to the public because of their erratic and consequently dangerous behavior. Tough little tubes about 5/16″ diameter with a "pulled in" choke were firmly charged with a fierce driver composition of gunpowder and charcoal (Fig. 79). When ignited they scurried along the ground and could be both amusing and irritating. They are, on the other hand, very effective in bags, mines and rockets, and are still used for this purpose.

Compositions:-

|                        | A  | B  | C  |
|------------------------|----|----|----|
| Meal gunpowder         | 64 | 91 | —  |
| Barium nitrate         | —  | —  | 60 |
| Potassium nitrate      | 8  | —  | 5  |
| Sulphur                | 4  | —  | —  |
| Charcoal 150 mesh      | 24 | 6  | —  |
| Aluminum "Bright"      | —  | 3  | —  |
| Aluminum dark pyro     | —  | —  | 25 |
| Aluminum 30/80 flitter | —  | —  | 10 |

Composition C requires a starter fire such as composition B

**Fig. 79**  Flying squib

*Cones*

Cones have been included in this chapter because they are really fountains, though they are not usually charged very solidly.

The cases are not too easy to roll, and are normally wet rolled, on a cone shaped brass or aluminum former, from a number of semi-circular pieces of paper cut out like Fig. 80.

The narrow end of the cone is then covered with touch paper or a plain paper circle pasted over the end, and they are placed narrow end downwards in a type of egg rack with holes large enough to support each cone in the center. The composition is poured into the cone, lightly consolidated by hand pressure, and then filled with sawdust to the top. Finally a disc is firmly glued in place or, in some cases, the bottom of the cone is spun over on to the disc.

Some cones are fitted with a washer inside the cone itself to give a secondary effect, but this is not common (Fig. 81).

Tough textile cones are quite effective for fireworks, though the writer is unimpressed by the huge cones seen on some markets, for a close examination reveals only about 2″ of composition at the narrow end!

Cone varieties are many and various, though as a general rule one is limited to the usual fountain effects – gold, silver, iron, gold and silver. Some cones commence their performance with colored fire, but this should always be perchlorate if it is to be followed by a mixture containing sulphur. In fact it is not even wise to mix perchlorate and sulphur, the articles get knocked about on the market and it is preferable to leave the cones as simply colored fire in this case.

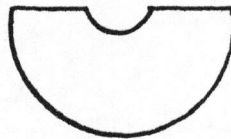

**Fig. 80**  Paper cut for rolling cones

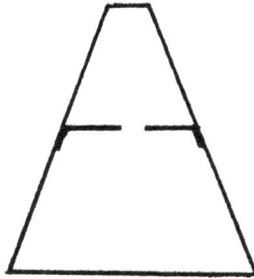

Fig. 81   Cone with secondary effect

Cone compositions are as follows, the one containing iron being taken from Weingart as it is a very satisfactory one.

|                              | A  | B  | C  | D  |
|------------------------------|----|----|----|----|
| Meal gunpowder               | 60 | —  | —  | —  |
| Charcoal 40/100 mesh         | 24 | 13 | 13 | —  |
| Titanium 40 mesh             | 16 | —  | —  | —  |
| Potassium nitrate 150 mesh   | —  | 54 | 52 | —  |
| Sulphur                      | —  | 9  | 10 | —  |
| Iron 60 mesh coated          | —  | 24 | 20 | —  |
| Charcoal 28 mesh             | —  | —  | 5  | —  |
| Potassium perchlorate        | —  | —  | —  | 63 |
| Aluminum "Bright"            | —  | —  | —  | 18 |
| Shellac 60 mesh              | —  | —  | —  | 9  |
| Aluminum flitter 30/80 mesh  | —  | —  | —  | 10 |

Composition D requires a starting fire and is usually damped with a 10% solution of shellac in alcohol.

# Chapter 14

# Pinwheels and Crackers

*Pinwheels*

There are several ways of making pinwheels, but good ones are not easy to manufacture. The usual British method employs a long narrow pipe about 3/16" in diameter, dry rolled against the grain, and pasted at the edge only. The length of the tube depends on the size of the pinwheel of course, but could be 12" to 20" long. The narrow bore tubes are not easy to fill, but they can be charged in bundles by funnel and wire or the composition can be shaken down the tubes. In England the wheels are filled on automatic funnel and wire machines operating on a bouncing principle.

If the composition is filled by hand bouncing, a sheet of paper is wrapped around a bundle of tubes so that it extends the open ends (one end of the tube having been previously closed with touch paper or sealed off in some other way). Fig. 82 will make this clear.

The composition is then poured on to the open ends of the tubes until they appear to be full. The whole bundle is then lifted and dropped sharply on to the bench several times, so that the composition is

**Fig. 82**  Bundle of tubes ready for charging

**Fig. 83**  Pin wheel

consolidated in the pipes. The whole operation is then repeated until the pipes are full.

Some manufacturers conduct this operation inside a container with a close fitting lid to reduce the dust hazard.

After charging, the pipes are placed inside damp cloths so that the paper will become sufficiently damp to avoid breakage during the coiling operation. When the tube is sufficiently damp it is neatly coiled around a cardboard or plastic disc, fixed with sealing wax or tape, and set aside to dry (Fig. 83).

Compositions are a little more complex than might be expected. In fact the mixures need to run down very narrow pipes and therefore contain a high percentage of grain gunpowder, coarsely ground sulphur and potassium nitrate crystals. They also need to be fast burning and they do contain a higher percentage of gunpowder than most printed formulations would seem to indicate. The following basic formulation is typical, but manufacturers naturally adjust the particle sizes of the ingredients according to their technique.

| | |
|---|---|
| Meal and grain gunpowder | 74 |
| Potassium nitrate | 13 |
| Sulphur | 13 |

Silver wheels can be made with gunpowder mixed with about 12% of bright aluminum powder.

Some silver spinning wheels of this type employ a mixture of barium nitrate, special kinds of dark pyro aluminum, potassium nitrate and aluminum flitter, similar to the silver gerb composition, but there are many snags to the manufacture of this firework. The composition is very hot of course and so it is essential to have a thicker and larger bore tube than the smaller pin wheels or else it will burn through all the coils of tube on ignition. In the second place, such a mixture is also virtually a flash composition and so it will explode if there are air locks in the filled tube; this can be partially overcome by rolling the

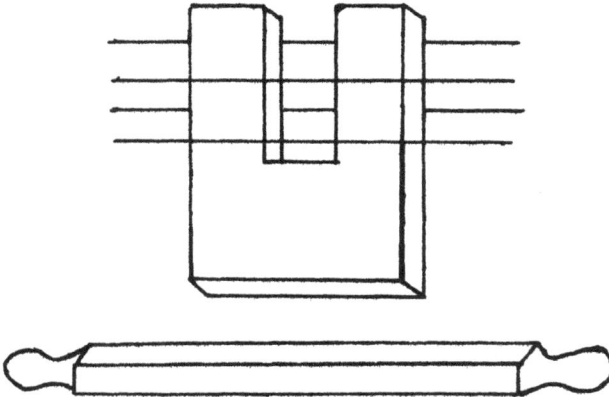

**Fig. 84**  Cracker bending tools

tube between heavy rollers after charging. A third drawback is that it would be dangerous to funnel and wire such a mixture, but it is very dirty to bounce it. The fourth problem is the type of aluminum to use. If the aluminum is too fast burning, explosion is likely, but if the aluminum is too slow the wheel will not turn. These wheels are a problem, but the Germans make good ones.

*Crackers*

English crackers are made in exactly the same way as pinwheels in that the grain gunpowder is poured over a bundle of pipes and then shaken down. The pipes are made of brown kraft paper, dry rolled. Manufacturers vary in their tastes as to which type of grain to employ but often use FFF or F or mixtures of these and other grades.

When the pipes are full of grain they are rolled several times between heavy rollers so that the grain is crushed and the pipes are flattened.

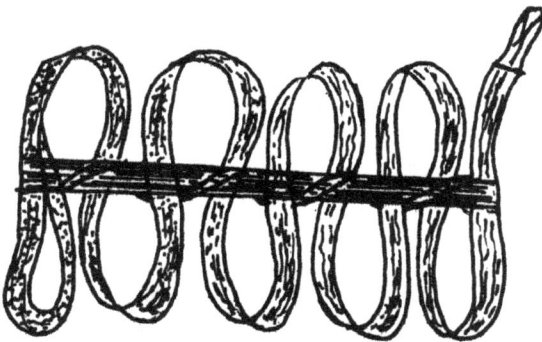

**Fig. 85**  Cracker

The extent to which the pipes are rolled is a matter of experience and can only be determined by trial and error.

After charging and rolling, the pipes are damped in the same manner as pinwheels so that they can be easily bent backwards and forwards into their characteristic form.

Weingart and others have described the method by which the pipes are bent backwards and forwards as in Fig. 84.

Finally the tubes are held in place by a tie of twine or adhesive tape (Fig. 85).

Some manufacturers use machines to bend the pipes, but others feel that the time taken to load the machine is almost as long as it takes to bend the pipes manually.

Large bore English crackers are rolled and charged on a manually operated machine which is fed with a continuous strip of paper and a hopper of gunpowder. A continuous length of charged cracker piping can be made in this way.

Many European manufacturers use an entirely different method for making crackers. The method is roughly as follows. The long narrow strip of paper for the tube is laid on the bench and pasted down one of the long sides; in fact several sheets are usually fanned out and pasted at the same time. The sheet of paper is then dry-rolled around a cylindrical brass former, but only half the paper is rolled on to the former. At this stage a specially long narrow scoopful of meal gunpowder is carefully placed onto the paper alongside the paper-covered former and the tube is finally rolled up and stuck down at the edge. The special gunpowder scoop naturally needs to be the same length as the cracker pipe itself. Finally, the former is removed from the pipe which is then

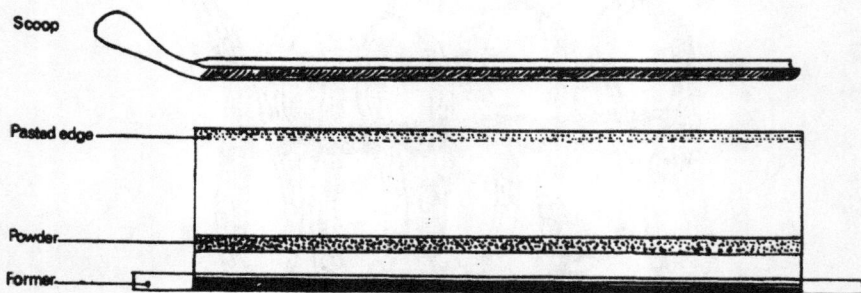

Fig. 86   German method of cracker manufacture

rolled and completed in the usual manner. The tools are shown in Fig. 86.

The narrow bore English crackers and the other wider bore crackers are somewhat different in performance. The former type tends to fizz in straight portions of the tube and then crack at the bends in a rather leisurely manner, but the wide bore crackers usually produce a quick succession of cracks rather like a burst of machine gun fire.

A rather cheap and less successful method of making crackers is to enclose a length of thick, good quality match in a close fitting length of match pipe, and complete it in the usual way.

*Torpedoes, Throw Down Crackers and Amorces*

These specialist items are outside the normal range of firework manufacture for they employ impact sensitive materials which are very very dangerous to handle. Silver fulminate is much too dangerous to handle commercially and mixtures of potassium chlorate and red phosphorus are quite lethal. Although the quantities used in amorces and torpedoes are quite minute and safe to handle, the actual process of manufacture is frighteningly dangerous and should only be attempted by the specialists who have problems of their own in any case.

# Chapter 15

# Indoor Fireworks

The manufacture of indoor fireworks tends to be somewhat specialized and is frequently undertaken by manufacturers who restrict themselves to these alone. They also have the advantage of being able to operate under less restrictive legal provisions.

The following represents a brief summary of the more usual types of indoor firework, but this sphere of activity is outside the direct experience of the writer who can do little more than comment in certain cases on confidential industrial information.

*Fern Paper*

This unusual effect is obtained by soaking brown ribbed kraft paper in hot water containing a mixture consisting of 5% potassium bichromate and 2 1/2% potassium nitrate. When the paper is dry, a sheet about 3″ by 4″ is folded backwards and forwards like the bellows of a concertina and fastened at one end to a cork or block of wood. When the paper is touched at the top with a burning cigarette, it smolders away to leave a beautiful and unusual fern-like ash. (Fig. 87).

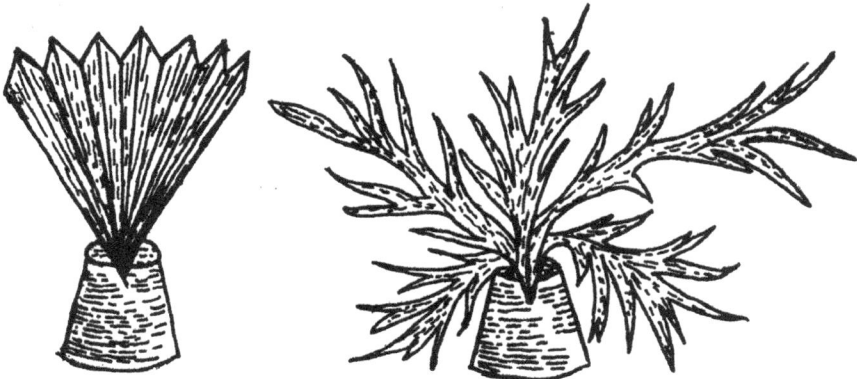

Fig. 87   Fern paper

173

### Fire Pellets

A compressed tablet of metaldehyde or hexamine will burn with a steady smokeless flame. The addition of lithium, or copper in some form, and magnesium powder will create quite attractive variations in color and effect. It should be noted, nevertheless, that the fumes produced by burning metaldehyde are objectionable in quantity, and in fact poisonous. Patents have been taken out for these fireworks in the past.

### Fire Pictures

Dr. Shimizu in chapter 19 has also mentioned these items. A sheet of semi-absorbent paper can be used for the picture which is drawn with a solution of potassium nitrate. The result is invisible when dry, but if any part of the picture is touched with a burning cigarette, the rest of the picture is gradually burnt out. Sometimes the papers are also printed in ink with additional detail. For example a man may be depicted firing a gun. The trajectory of the bullet is invisible until it is burnt on the saltpeter trail, and it culminates in the ignition of a paper amorce fixed at the end.

### Flash Paper

Sometimes unsized paper is nitrated with the usual nitrating mixture of concentrated nitric and sulphuric acids. When it is thoroughly washed and dried, it burns with a brilliant flash. It is well known though that such nitrations need careful control and the products can be very unstable unless they are carefully washed and expertly made.

### Fountains

Indoor fountains are almost odorless and smokeless, being made principally from nitro-cellulose. The addition of iron, magnesium and aluminum along with salts of lithium and copper produces some very pleasant effects. The main problem is knowing where to obtain powdered nitro-cellulose with the correct percentage of nitrogen and the correct, and very necessary, stabilizer. Sometimes when the conditions are not right, the fountains produce gas but no flame.

### Matches

The manufacture of match heads has been well covered elsewhere (5), but the so-called Bengal Matches do come within the scope of fireworks even though they are normally made by the match manufacturers. It is well known of course that the choice of adhesives, the viscosity and the temperature of the slurry are all important when the splints are dipped, for this affects the way in which the matches dry. Some pre-war German formulations set out in the B.I.O.S. Reports (12) were as follows:-

| Red | Strontium nitrate | 2,500 |
|---|---|---|
| | Shellac | 500 |
| | Potassium chlorate | 300 |
| | Fine charcoal | 250 |
| Green | Barium nitrate | 3,000 |
| | Shellac | 500 |
| | Potassium chlorate | 300 |
| White | Potassium nitrate | 3,500 |
| | Sulphur | 1,125 |
| | Shellac | 1,000 |
| Blue | Potassium chlorate | 900 |
| | Paris Green | 300 |
| | Resin (Colophony?) | 135 |

Storm matches are simply bengal matches producing flame but not color. They were used extensively at sea for the older signals which had to be ignited by ordinary methods, and are still used for lighting portfires at firework displays.

Exploding matches are usually book matches which have a tiny dab of fulminate and adhesive at the base of the match head so that they produce a sharp report soon after ignition.

*Smokes*

Firework smokes are frequently used in the theatre, but naturally they must be as innocuous as possible. Sometimes a smoke puff is produced by the combustion of a flash powder made with aluminum, magnesium or zinc; larger quantities of smoke can be made with the insecticidal types of smoke (see chapter 17).

Colored smokes are made with the usual combination of dye, potassium chlorate and sucrose or lactose. An interesting colored smoke stick can be made by forming the composition into a slurry with adhesive and dipping the sticks in the same way as for sparklers.

An amusing little non-pyrotechnic smoke seen on the market utilizes celluloid to give the illusion of a toy monkey smoking a cigarette.

*Snakes*

The best snakes are the black ones made out of pitch. This curious German discovery gives an immense amount of pleasure, for a pellet the size of a 5/8" roman candle star will produce an ash 1" in diameter and four feet long. The Germans still produce very good snakes.

Weingart and Davis (6,7) have written at some length on the manufacturing procedure and so there is no need to repeat this except to add a few of the caveats omitted by them.

The most successful snakes are made from naphthol pitch, the residue after the distillation of $\beta$-naphthol. This material is not readily available and the manufacturers naturally make no guarantees for what is a waste product; samples have to be obtained and tested. Davis maintains that a mixture of roofing pitch and syrian asphalt is a good substitute. Some manage to use ordinary coal tar pitch.

The actual nitration is an unpleasant procedure, best carried out in an aluminum or earthenware bowl. The reaction is very vigorous and copious clouds of nitrogen dioxide are evolved. Ordinary concentrated nitric acid will not work and fuming nitric acid is too strong; therefore a small amount of water should be added to the fuming nitric acid. This also enables the reaction to proceed reasonably smoothly.

Weingart and Davis do not make it clear that the oxidation process is incomplete. The addition of too much fuming nitric acid frequently causes the mass to catch fire, destroying the whole experiment. On the other hand insufficient acid for the completion of the reaction, causes the mass to be sticky due to the presence of too much oil. The correct quantity of acid leaves a hot molten mass with no excess acid present, due presumably to the usual breakdown of nitric acid in the presence of hot carbon. Washing is thus unnecessary, though as a precaution a

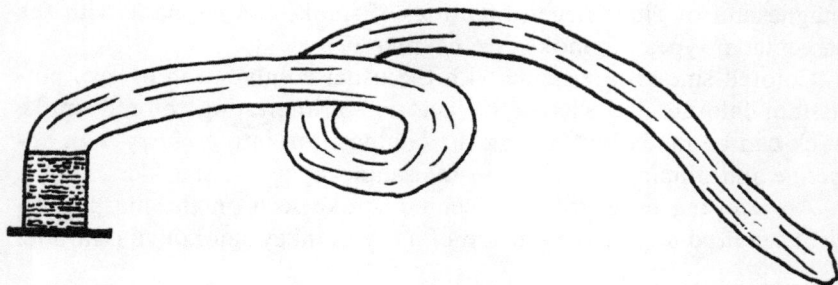

Fig. 88   Firework snake

**Fig. 89**    Snake-in-the-grass

tiny percentage of barium carbonate is sometimes incorporated into the final operation.

After the mass has cooled and solidified it is ground and pressed into pellets with the addition of a small amount of lubricant e.g. graphite. Picric acid can be used to burn the pellets for producing the ash, but it is very unsatisfactory commercially since it creeps to the surface, stains the fingers and clothing, and has a bitter taste. Nitrocellulose gives the best results and apparently the Japanese use ammonium perchlorate. (Fig. 88)

The snake-in-grass is an interesting variation. A small cone made of aluminum foil is filled with a mixture of ammonium dichromate and a small amount of fuel and oxidizer. At the base of this a pinch of loose black snake composition is added before the cone is sealed off. When the cone is ignited it produces a small spray of green chromic oxide, followed by the black snake. (Fig. 89).

The much smaller white snakes are made with mercuric thiocyanate $Hg(CNS)_2$. This obscure salt is insoluble in water and can be made by precipitation from solutions of potassium thiocyanate and mercuric chloride. When the precipitate has been washed and dried it is made up into tiny pellets with a solution of gum arabic and potassium nitrate in water. This snake is much less exciting than the black one and has the disadvantage of being poisonous.

*Snow Cones*

These little cones are made in exactly the same manner as the snake-in-the grass. Circles of aluminum foil are cut out with a radial cut as shown in Fig. 90, and after forming into cones they are placed in a

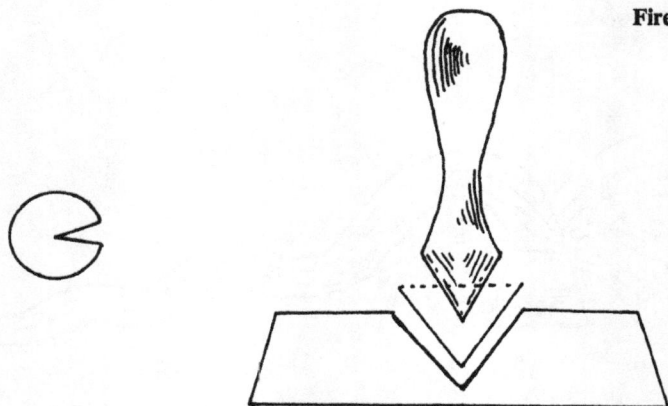

**Fig. 90**   Tools for snow cones

board with a series of cone shaped holes of the same size (Fig. 90). When the cones are filled in this way, they retain their shape and are easily closed up at the base by simply bending the foil over on top of the composition.

The snow cone merely burns magnesium powder with the production of large flocks of white magnesium oxide, though quite why any house-wife should wish to allow such a thing to happen indoors is anybody's guess. The composition is basically;-

| | |
|---|---|
| Fine magnesium powder | 40-50% |
| Potassium nitrate | 5-10% |
| Flour or woodmeal | about   40% |

Small tablets of metaldehyde also produce a snow-like effect when placed on a cigarette due to the sublimation of the metaldehyde.

*Sparklers and Fire Sticks*

The well known sparkler is basically made of barium nitrate, alumi-num and steel.

| | |
|---|---|
| Barium nitrate | 50 |
| Dextrin | 10 |
| Steel | 30 |
| Aluminum powder | 8 |
| Charcoal 150 mesh | 1/2 |
| Neutralizer | 1 1/2 |

The materials are ground with suitable adhesives and water to make a thick slurry. Wires are fixed into wooden frames to space them suit-ably before dipping into the slurry. Sometimes the sparklers have to be dipped twice. As might be expected, the drying process is quite im-

portant and usually takes quite a long time to prevent the mass cracking. The iron is usually coated in good quality sparklers, though
judging by the appearance of many commercial products, this is frequently not the case.

Other fire sticks are made also, but of course they should not be
burnt indoors as the compositions suggest. The following composition
is from an earlier American patent number 1, 936,221 (1933)

| | |
|---|---|
| Potassium chlorate | 46 |
| Barium nitrate | 17 |
| Strontium carbonate | 12 |
| Shellac | 11 |
| Cryolite | 8 |
| Dextrin | 6 |

The outside of the composition is coated with magnesium/aluminum
alloy grit. Weingart gives other compositions but these have not been
tested by the writer.

*Table Bombs*

Table bombs are in effect little mines filled with toys and other items.
The lid is fixed quite lightly so that it will not blow off violently, and

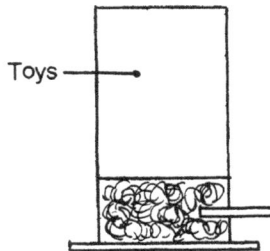

**Fig. 91** Table bomb

instead of blackpowder, a nitrated cotton is used to eject the toys. The cotton is ignited with Bickford fuse.

## Theater Fires

Smokeless fires are sometimes required, but it is not possible to produce good colors without smoke or smell. The following compositions are sometimes used, but they must be kept dry.

| | | | |
|---|---|---|---|
| Strontium nitrate | 68 | Barium nitrate | 56 |
| Potassium chlorate | 24 | Potassium chlorate | 26 |
| Shellac 30/120 mesh | 8 | Shellac 30/120 mesh | 18 |

# Chapter 16

# Fuses, Quickmatch

## Fuses

*Safety Fuse*

The so-called Bickford fuse was invented by William Bickford in 1831 to overcome the somewhat erratic methods of igniting blasting powder. Fordham (13) describes how the fuse is made from a black-powder core which is surrounded by jute yarn and subsequently coated with bitumen on the outside. A typical formula for the blackpowder grains would be

| | |
|---|---|
| Potassium nitrate | 65 |
| Sulphur | 24 |
| Fine charcoal | 11 |

In Great Britain the fuse has to burn within certain limits of 80 to 100 seconds per yard, though variations in burning speeds would naturally be expected under varying conditions.

The firework maker primarily requires fairly short lengths of fuse which provide delay times up to about 10 seconds at the most. Bickford serves this purpose well and has the additional advantages that it

Fig. 92  Bickford cut and primed

181

**Fig. 93**  Bickford threaded with quickmatch

is reasonably waterproof, the fire is not communicated laterally, it is easily cut and pierced, and is tough enough to block up narrow bore tubes in addition to acting as a delay.

Bickford made by the Imperial Chemical Industries is normally about 3/16″ in diameter. This narrow bore is commonly used in small explosive fireworks and ignites quite easily from a gunpowder charge. To ensure ignition the fuse is often cut at 45° and then primed with gunpowder paste. (Fig. 92).

The best way to be certain of ignition is to punch a hole through the center of the Bickford from one side to the other and then thread this with a length of quickmatch. The small diameter of the British product does not readily lend itself to this technique, but the larger 3/8″ sizes made in other parts of the world are excellent for this purpose. (Fig. 93).

In addition to making the larger size of Bickford, the Americans also make a self-consuming type of Bickford fuse which is useful for firework purposes where it is intended to ignite a succession of items. This Red Visco, as it is called, has only a small amount of yarn on the outside and is finally coated with a layer of red nitrocellulose dope.

More recent times have seen the introduction of plastic Igniter Cord. Again according to Fordham (13) the fast cord burns at a rate of about 1 second per foot and consists of paper or textile yarns coated with blackpowder at the center. This is then covered with a plastic incendiary composition beneath an external coat of waterproof polyethylene. The slow cord is made in a similar manner except that a copper wire replaces the blackpowder center of the fast cord. The copper wire conducts heat from the burning front into the unburnt composition thereby controlling and speeding up the rate of burning. The slow cord burns at the rate of about 10 seconds per foot. The plastic incendiary compositions are usually made from potassium nitrate or pot-

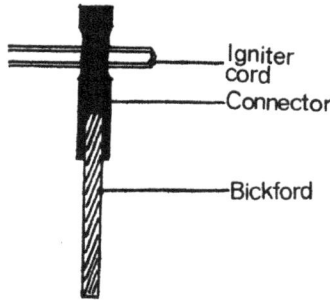

Fig. 94   Method of connecting several pieces of Bickford fuse

assium perchlorate, red lead and silicon with a nitrocellulose binder. At the time of writing, the comparatively new igniter cords are cheaper than piped match and have the advantage of being waterproof. On the other hand they are much slower than good piped match and in long lengths burn irregularly, appearing to hang fire momentarily in an unusual manner. A further disadvantage of igniter cord is that it cannot be easily used for lancework and is more likely to be accidentally ignited from dross than paper matchpipe. However igniter cord is very useful for igniting a number of lengths of Bickford by using the special Bean Hole Connectors which are also made by I.C.I. The technique is shown in Fig. 94.

*Firework Fuses*

Firework makers sometimes make their own time fuses with blackpowder. Narrow-bore tubes with strong walls are charged with fine grain powder. Each increment is rammed with about twelve blows of a mallet and there are about eight increments per inch. Much, of course, depends on the powder that is used, but such fuses usually burn at the rate of about three seconds per inch. These fuses can also be pressed.

Occasionally fuses are made with modified gunpowder mixtures e.g.

|                     |    |
|---------------------|----|
| Potassium nitrate   | 25 |
| Sulphur             | 10 |
| Meal gunpowder      | 65 |

Fuses for small mines and cannon crackers are long narrow bore tubes charged with a reasonably fast composition of the golden rain type e.g.

| | |
|---|---|
| Meal gunpowder | 75 |
| Potassium nitrate | 20 |
| Sulphur | 5 |

The composition is charged with a funnel and wire.

### Quickmatch

One of the most essential items to the firework maker, quickmatch, is also one of the most unpleasant to make. Cotton strands are passed through a slurry of gunpowder, adhesive and water and then wound on to a drying frame and allowed to dry.

The match is usually made up from four to ten strands of cotton which are run off reels into a bath of alcohol and water, which thoroughly wets the cotton. As the cotton leaves the bath it passes between rollers which remove the alcohol solution prior to passing into the gunpowder slurry. As the cotton leaves the slurry it passes through a small funnel which removes the excess.

The match is wound on to large wooden frames, about six feet long and four feet wide, where it stays until it is dry. Frequently the box of slurry moves along a threaded rod which in turn is geared to the winding frame so that the match is carefully spaced on the frame as it leaves the slurry container.

When the frame is full of match it is dusted with fine grain powder while it is still wet. Manufacturers use a good quality powder for match, (e.g.FFF) but they vary in their choice of adhesives. European firework makers tend to favor gum arabic or dextrin, but in Britain there is a preference for boiled starch solution. Perhaps the damp British climate is the reason for this preference for there is no doubt that match made with dextrin becomes quite limp in damp weather.

A typical match formulation would consist of approximately 2 1/2 parts of grain powder to 1 part of a 20% solution of starch.

Good match burns steadily and fiercely at about the rate of 5 seconds per foot in the open. On the other hand, when placed in a match pipe it is almost instantaneous.

Match pipes are made from brown kraft or white sulphite paper. The tubes are dry rolled on a former about 1/4" in diameter; they must be dry rolled since they are bent so much when in use, but the diameter of the pipe is not especially important. Some manufacturers make their match pipe in continuous lengths from rolls of strip paper, but there is still a tendency in Britain to roll the pipes in 20″ lengths and join them together as required. For this purpose, the former is slightly tapered so that the narrow end of one pipe will fit into the wide end of the next.

# Chapter 17

# Smoke

The extensive use of smoke for screening and signalling purposes during the last two World Wars has produced a very specialized branch of pyrotechnics which is not directly concerned with fireworks. As there is an abundance of specialized literature on the subject, the reader who is looking for greater detail is referred to the following works (4, 5, 9, 12, 22).

Colored smokes are usually made by vaporizing a suitable dye by mixing it with a heating mixture of potassium chlorate and a fuel. In order to make the mixture as cool burning as possible, lactose or a mixture of lactose and sucrose is used as the fuel. In certain types of white smokes thiourea is also sometimes used.

As a general rule with colored smoke formulations it is necessary to use dyes which have a melting point which is as consistent and as low as possible. The dye should also be free from organic salts and have a particle size which is uniform. Smoke compositions are mixed for long periods to produce uniformity and consistency, for this, along with the particle size, affects the burning rate which is further controlled with other additives.

A simple smoke formulation would consist of approximately 60%

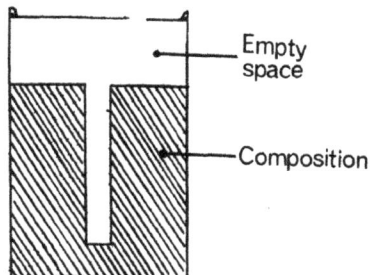

Empty space

Composition

**Fig. 95**  Smoke cannister

dye, 20% potassium chlorate and 20% sucrose or lactose or a mixture of the two. The material is charged into a metal container with hydraulic pressure, leaving a space down the center of the composition rather like the spindle hole in a rocket. The composition must not be too dense or voluminous or else the carbonaceous residue will decompose the smoke before it leaves the container. A metal gauze is frequently placed on top of the smoke composition, which must not completely fill the cannister. The cannister is then closed with a suitable lid which contains holes for the smoke to issue forth, and a suitable ignition. The empty space at the top of the container is essential in order to allow cooling to take place. In the absence of this, the smoke will catch fire and merely burn. These smoke containers and ignitions have become highly complex and efficient. A smoke cannister is shown in Fig. 95.

Black smokes gain their effect from the production of tiny particles of carbon when oxygen negative mixtures containing naphthalene or anthracene are burnt. These mixtures consist of potassium perchlorate and anthracene or potassium perchlorate, hexachloroethane and naphthalene etc. During the last war the Germans used a somewhat explosive mixture of hexachloroethane, anthracene and magnesium powder e.g.

| | |
|---|---|
| Hexachloroethane | 60 |
| Anthracene | 20 |
| Magnesium-fine | 20 |

Brown smokes have frequently been made with pitch or tar. Hard pitch is ground to a powder and mixed with potassium nitrate and sulphur. Glue is frequently added to control the reaction. The mixture is placed in a cannister similar to colored smoke, but it tends to get very hot. The following composition given by Faber (9) is typical:-

| | |
|---|---|
| Pitch | 29.2 |
| Potassium nitrate | 47.4 |
| Borax | 10.6 |
| Calcium carbonate | 4.9 |
| Sand | 4 |
| Sulphur | 3.9 |

A better method is to use liquid tar which is absorbed into sawdust. The mixture has better burning characteristics and can be used in thick-walled paper tubes.

Grey smokes are invariably of the so-called HC type involving the use of hexachloroethane (HC) and zinc in some form. The grey color is formed because of the production of zinc chloride and carbon particles, but the zinc chloride also forms zinc hydroxide and hydrochloric acid in the presence of moisture. Mixtures containing zinc powder are liable to react with water and it is usual practice to make sure that moisture is excluded before the cannister is sealed. One magazine was destroyed when smoke spontaneously ignited because a worker allowed beads of perspiration to fall into the composition during processing. In view of this hazard, HC smokes are more frequently made with a mixture of hexachloroethane and zinc oxide with smaller percentages of calcium silicide and potassium nitrate. The high vapor pressure of hexachloroethane also means that all the HC smokes have to be sealed in air-tight cannisters.

|   | | |
|---|---|---|
| A. | Hexachloroethane | 50 |
|    | Zinc Powder | 25 |
|    | Zinc oxide | 10 |
|    | Potassium nitrate | 10 |
|    | Colophony resin | 5 |
| B. | Hexachloroethane | 45.5 |
|    | Zinc oxide | 45.5 |
|    | Calcium silicide | 9 |

A is from Izzo (8) and B is from Ellern (5).

White smokes were invariably made by the old firework makers with ammonium chloride. Strictly speaking, a mixture of potassium chlorate and ammonium chloride is highly unstable owing to the formation of ammonium chlorate. Nevertheless there seems to be good evidence of the stability of such mixtures. Shidlovsky quotes the following:-

| | |
|---|---|
| Potassium chlorate | 20 |
| Ammonium chloride | 50 |
| Naphthalene | 20 |
| Charcoal | 10 |

and Dr. Becher (10) the following:-

| | |
|---|---|
| Potassium chlorate | 40 |
| Ammonium chloride | 45 |
| Montan wax | 12 |
| Kieselguhr | 3 |

Formulations of the latter type occasionally employ a little paraffin oil presumably to help exclude moisture and reduce sensitivity.

More useful white smokes are made with HC and zinc oxide or mixtures of potassium chlorate, thiourea, lactose and the organic chemicals used as insecticides.

The earliest types of insecticidal smoke generators were used in the second World War in the form of hand grenades for fumigating dugouts in the jungle and the Far East. Later these were developed, under patent, for use in greenhouses.

Insecticidal smoke formulations are much the same as those for colored smokes which utilize dyestuffs, though in this case the pyrotechnic mixture has to be carefully regulated to give the right amount of heat to vaporize the maximum quantity of active ingredient.

The earliest ingredients in these smokes were DDT and BHC

DDT 1,1,1-Trichloro 2,2-D1-(4-chlorophenyl) ethane

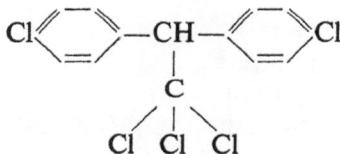

$$Cl-\langle\ \rangle-CH-\langle\ \rangle-Cl$$
$$\underset{\underset{Cl\ \ Cl\ \ Cl}{|}}{C}$$

Lindane (gamma BHC) $\gamma$-1,2,3,4,5,6-Hexachlorocyclohexane, the pure gamma isomer later replaced BHC. Later developments have also utilized azobenzene and tecnazene (1,2,4,5,-tetrachloro-3-nitrobenzene).

After ignition of the smoke composition, a dense smoke is produced with particles less than $3\mu$ in diameter which consist of super-cooled droplets of insecticide. As the insects fly around (particularly at optimum temperatures of 21 °C.) they collect the insecticide particles on the hairs of their bodies and antennae and soon die.

Other firework smokes have been discussed by Dr. Shimizu in chapter 19. Ordinarily, smoke has no direct application in fireworks; on the contrary it can be a nuisance at a firework display on a still night. On the other hand, apart from daylight fireworks, smoke is used commercially because of its penetrating capacity. It is used to detect leaks in drains and tanks, for it not only penetrates all available space, but is visible when it emerges.

Drain testers are simply thick-walled tubes clayed like fountains but with a smaller hole, and charged with a simple mixture of potassium nitrate, sulphur and a small amount of additional fuel. These smokes contain large amounts of sulphur dioxide and are also frequently used to discourage mice and moles. Mole smokes are usually made of roughly equal quantities of potassium nitrate, sulphur and sawdust. They do not often kill the animals, but drive them into the next property instead!

# Chapter 18

# Exhibition Fireworks

Exhibition fireworks do not vary very much from one manufacturer to another in their general arrangements because the number of possibilities are limited. On the other hand the finer details vary enormously from one country to another though we can see that each individual country tends to follow its own general pattern.

In England displays are usually fairly leisurely affairs striking a balance between set ground pieces, roman candles, rockets and shells. On the other hand the Italians and Germans prefer fast firing mass aerial effects of shorter but more spectacular duration. Many English people also prefer the latter type of display but the public usually feel that a show should last half an hour if they are to get good value for money. Apparently also the English are more reluctant to spend large sums of money on displays, preferring to buy a few smaller fireworks

**Fig. 96**  Mortar boxes at Pains-Wessex, Ltd.

of their own. Large displays are still very much a feature of European fetes, festivals and carnivals, but they are becoming more rare in England. Very large quantities of small fireworks are still sold in England for the traditional festival on November 5th (Guy Fawkes Day) possibly to a value of about five million pounds sterling.

*Shells*

Shells for display work have been fairly adequately covered in chapter 12. Re-loading is virtually impossible in fast firing displays making it quite imperative to have one mortar per shell. For the smaller sizes, paper mortars in wooden boxes surrounded by sand are quite adequate, but care must be taken not to allow the boxes to fall over. The shells are frequently fitted with automatic delays of the type mentioned earlier, but if these are used, the match pipe must be tied to the mortar or else a shell leaving a mortar can sometimes drag unfired shells after it. (Fig. 96)

*Roman Candles*

Display candles are usually fired in batteries of five or more. The best effect is gained by arranging the candles in a fan-shaped board or spline. A piece of timber drilled with a few holes to take the tubes is all that is required. (Fig. 97)
Candles are sometimes fired in bundles but these are less effective.

At this stage it would be appropriate to mention the manner in which candles and other fireworks are joined together with piped match. The candles and gerbs are usually capped with about three turns of a tough thin paper (e.g. ribbed kraft). The paper is pasted only at the edge and

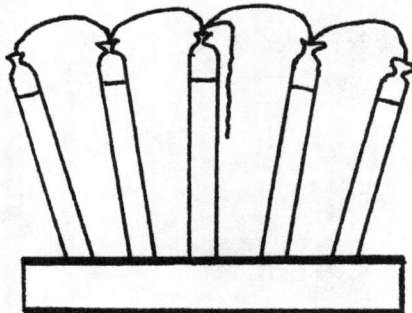

**Fig. 97**   Roman candle battery

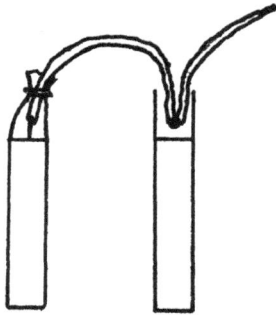

Fig. 98  Matching tubes

where it comes into contact with the tube. It should overlap the end of the tube by about 1 1/2".

When it is necessary to join several tubes together, a length of piped match is used to connect them. The end of the match pipe with about 1/2" of bare match protruding is pushed inside the end of the tube and the capping paper gathered around it. The capping paper is then tied with a clove hitch. The match pipe is then cut further along the pipe near to the next tube with a V shaped snip to bare the match again, and after being inserted into the capping it is tied off as before. The method is illustrated in Fig. 98.

The clove hitches are almost always used by fireworkers and are tied in two ways (Fig. 99). Both methods have to be mastered, for the one method involving the formation of two loops on top of each other, can only be used in those situations where it is possible to slip the loops over the end of the object to be tied. It is also important to note that when the cappings are tied they must be neither too tight nor too slack, for in the former situation the fire does not always get through, and in the latter case the match can be pulled or blown out of the cap without causing ignition.

Fig. 99  Clove hitch

**Fig. 100**   Collapsible rocket frame

*Rockets*

Large display rockets are fired from frames with a double row of screw-eye staples about 3″ apart. The frame is usually collapsible for ease of transportation and arranged so that the rockets can be fired at any angle. (Fig. 100)

**Fig. 101**   Rockets arranged to fire in quick succession

For this type of firing, the base of the rocket is covered with a paper drum-head so that the sparks do not ignite other rockets in the frame. Touch-papered rockets cannot, of course, be fixed in these frames. Sometimes the rockets are furnished with thin cappings and matched together for rapid firing.

Mass firing of small rockets is obtained by placing many rockets on to a board drilled with holes, or chicken wire covered with a sheet of paper, or simply in long rows. Each rocket has a piece of match in the vent and in each case ignition is obtained from one of the rockets which produces enough sparks to light the rest. Batteries of rockets of this type require some kind of cover lest they ignite prematurely owing to stray sparks from other items in the display. (Fig. 101)

Falling rocket sticks are a potential hazard, of course, but it is remarkable how few seem to cause any damage, particularly to people. It is also necessary to exercise some caution in firing parachute rockets, since chutes occasionally fail and drop burning flares to the ground.

*Fire Pictures*

Prior to the 1939-45 war lance-work was very much a feature of firework displays. Huge portraits of Kings and Queens 20 feet by 30 feet, Indian Palaces, Triumphal Arches were to be found, involving the use of thousands of lances and many more feet of match. The assembly of these items involved large quantities of timber, posts, ropes and pulleys as well as very many man hours. A whole day or more would be spent setting up these displays with rain clouds hovering menacingly above. Needless to say such displays are now most uncommon; four or five frames 10 feet by 5 feet is the most one would expect to see.

Lance-frames are usually a standard size of 10 feet by 5 feet in Britain, being made in one foot squares from light laths. The portrait or

Fig. 102   Making lancework frames

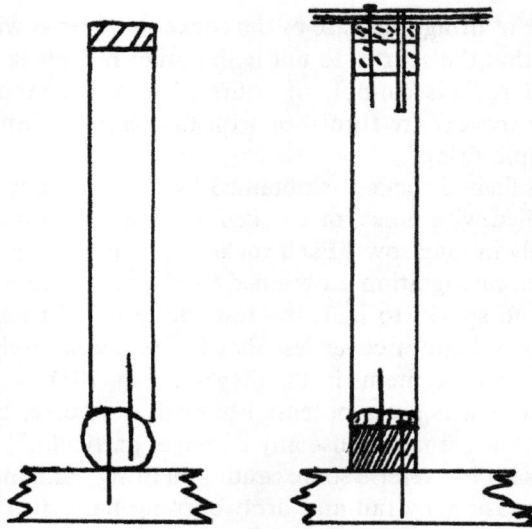

**Fig. 103**   Matching lances

design is first drawn out on squared paper or graph paper and then reproduced exactly on the 10 by 5 frame using thick cane which is fastened to the frame with wire nails.

The next stage is to fix double pointed nails into the canework at 3″ intervals. These nails are about 1/2″ long and are allowed to project about 1/4″ into the cane. (Fig. 102).

The unprimed ends of the lances are dipped in glue and then pushed

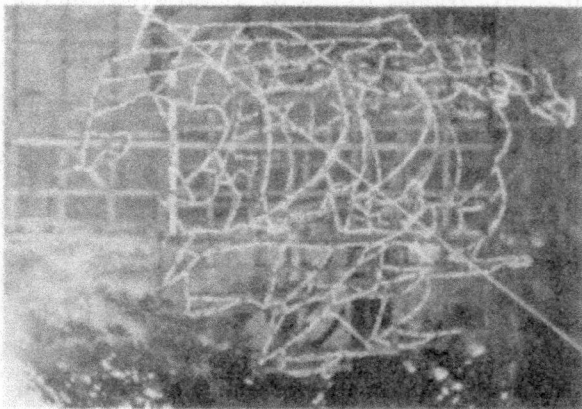

**Fig. 104**   Ship in lancework

on to the nails, where the glue sets sufficiently to hold the lances temporarily.

Techniques for connecting the lances with piped match vary somewhat but are all more or less similar. The most laborious method uses a large pin to secure raw quickmatch to the lance. This is then covered between each lance with short lengths of piping which are slid along the match. Pasted paper strip is then used to cover the top of each lance. This method is ridiculously time-consuming but it does ensure that all the lances ignite. (Fig. 103).

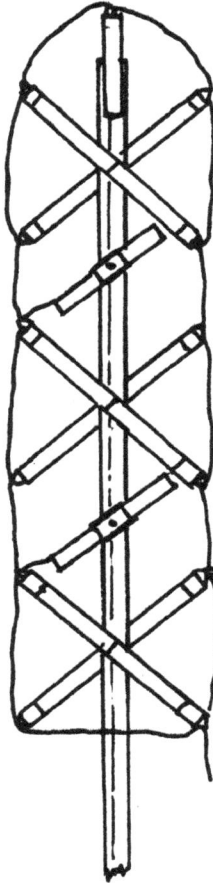

**Fig. 105**   Lattice pole before ignition

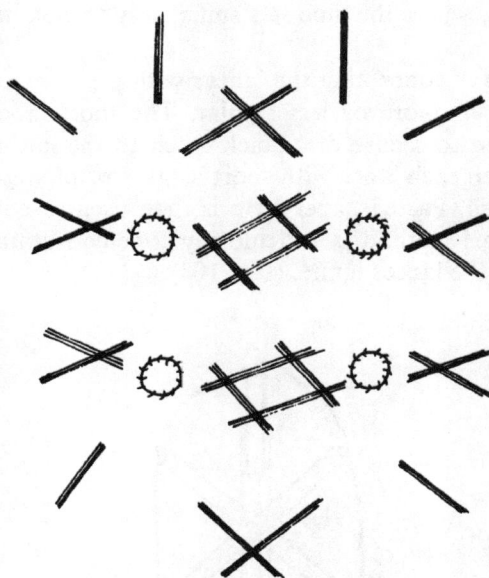

Fig. 106   Effect of two lattice poles ignited

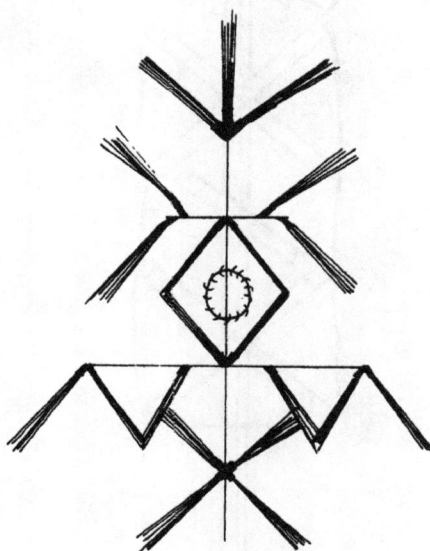

Fig. 107   Gerbs forming a tree piece

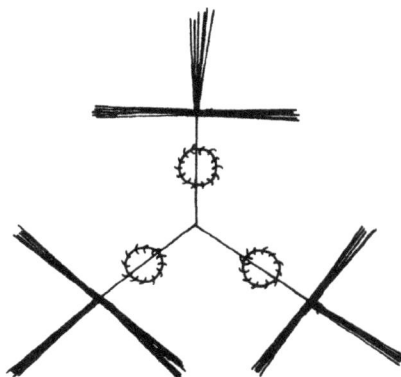

Fig. 108   Gerbs and saxons in geometric design

An easier method is merely to lay piped match across the tops of the lances and drive a large pin through the pipe into the lance. A narrow staple is also used for this purpose. It is also fairly common practice to make an additional hole through the match pipe into the lance prime, to ensure the fire transfer from the match. The final operation is to cover the top of the lance and the match pipe with either pasted paper or adhesive tape. This operation helps to prevent the pins from being pulled out and is a protection from rain and stray sparks. The lancework can also be protected from rain with a solution of shellac or paraffin wax. A ship in lancework is shown in Fig. 104.

*Fountains*

Fierce burning fountains or "Fix" are arranged on light timber to produce geometrical designs or "feux croisées". In situations where it is desirable for fires to touch or cross, the spread of fire needs to be measured so that distances can be worked out. Saxons and small wheels are invariably added to produce variety and movement. It is here where the greatest variations are to be found, each fireworker being able to devise his own pieces. The following suggestions are typical. (Fig. 105-108).

*Wheels*

Display wheels vary in size from 2 feet, up to 15 feet in diameter. The smaller wheels (up to about 5 feet) are frequently made in the old fashioned manner with a heavy central boss, six spokes and a circular beech or ash hoop around the circumference.

Larger wheels are made so that they can be assembled on the site.

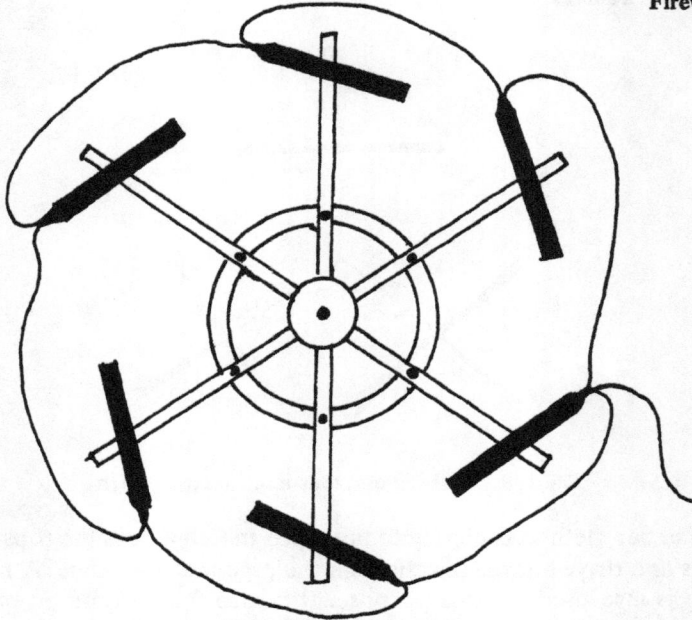

**Fig. 109**  Revolving sun

In this case, the spokes slot into the boss in tapered sockets and are held in place by a smaller steel hoop and a number of bolts. (Fig. 109).

Many horizontal and vertical wheels are made from a long piece of timber with a hole drilled through the center to take the spindle upon which the timber revolves. (Fig. 110).

Vertical wheels are often fitted with color pots, whistles, gerbs and even waterfall cases, strings of lights or flash charges. (Fig. 111).

Drivers for wheels vary very much with the size and speed of turning. Small wheels about 3 feet in diameter use 3/4" drivers; up to 10 feet take 1" drivers; 15 foot wheels take large gerbs about 1 1/2" internal diameter and 10 inches long.

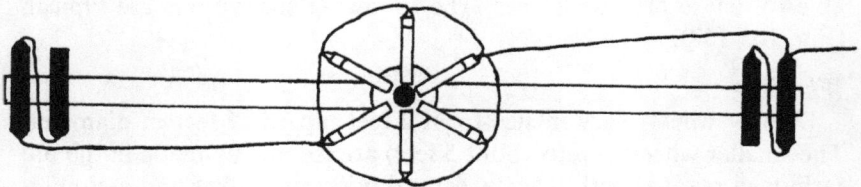

**Fig. 110**  Large vertical wheel

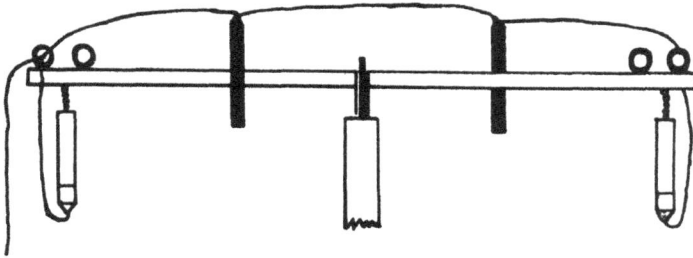

Fig. 111   Horizontal wheel

*Combination Fronts*

Very large spectacular effects are obtained by combining several identical pieces or a mixture of set pieces and wheels etc. (Fig. 112).

Battle scenes also use various combinations. Two ships in lance-work for example can be made to fire streamer roman candle stars at each other to the accompaniment of roman candle whistles, flash charges, thunderflashes and crackers. The possibilities are quite endless.

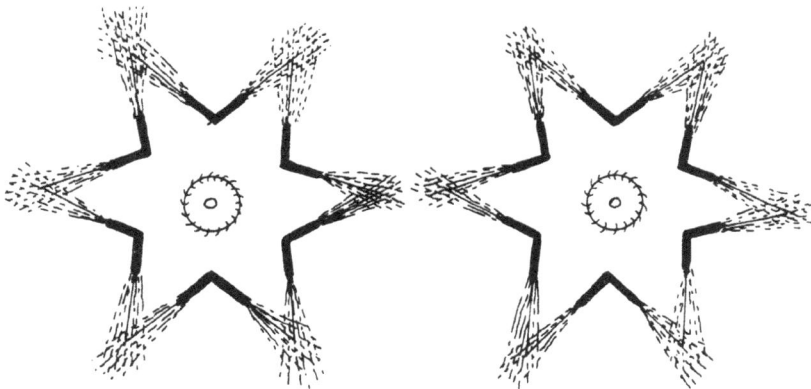

Fig. 112   Combination fronts

# Chapter 19

# The Manufacturing Processes of Firework Composition

## Japanese Fireworks
## Takeo Shimizu

These processes are generally very simple and, having learnt how to manufacture the most dangerous compositions, it is easy to manufacture others which are safer. The compositions which contain potassium chlorate and red phosphorus, realgar, antimony trisulphide, metal powder are the most dangerous at present. The best way is to use none of them, but we cannot always find other safer compositions of the same character as these, and we often have no alternative but to use them.

We have two processes, the wet and the dry.

The *wet process* applies to compositions of high sensitivity, which do not explode in the wet state but explode when the moisture is driven out by drying. The wet process is, for example, as follows: A quantity of water is added to an oxidizer and mixed into a wet state. Then some fuel is added to it and the whole is well mixed to a paste, keeping the wet state. Compositions of potassium chlorate and realgar, or potassium chlorate and red phosphorus, are usually produced by this process. When the process is finished, we must use it quickly by wrapping it in small pieces of paper or by some other method of separating the composition into small amounts before drying. The composition which is enveloped in each piece is left in the air so that it may be dried gradually and gain its sensitive character.

It must be noted that, in the case of the wet process, it does not apply to ingredients of high solubility or to the ingredients which cause some chemical reaction or create some acid in the water. We have had experiences of some spontaneous explosions which are thought to have been caused because the acid concentration became high enough to cause the explosion. Each component has a different solubility in water, and when the wet composition is left to stand for a while, some particles of the ingredients which are heavier than others sink to the bottom, and cause a lack of uniformity. An adequate amount of water and if necessary, gelatine, can be used to moderate the viscosity of the composition.

The minimum amount of water, which ensures safety during mixing, depends upon the composition, for in a composition which contains red phosphorus and potassium chlorate 30 grams of water per 100 grams of the composition will not suppress its inflammability and explosiveness perfectly; in fact the amount of water should not be less than 50 grams. A water content as high as this makes the composition muddy. A composition, which contains potassium chlorate and realgar, sulphur or antimony trisulphide, explodes even when mixed with 20 grams of water per 100 grams of the composition, but becomes non-explosive with 30 grams of water per 100 grams, so in this case the operation is permissible with not less than 40 grams of water. Care must be taken during the operation that neither the rim of the container nor one's clothes are stained with the slurry which can often dry and ignite without one being aware of it.

The *dry process* applies to compositions which are relatively safe to handle. This is to mix dried powdered ingredients, in a dry state, by means of sieving, and this does not change the component ratios like the wet process. (The wet process causes the oxidizer to ooze out of products thus changing the ratios of the components). First the ingredients are ground into fine powder with a chemist's mortar, wooden roller or ball mill. It is one of the common rules to sieve each ingredient before mixing, so that the particle size is uniform and foreign substances are eliminated. In the case of compositions of relatively high danger like a potassium chlorate or barium chlorate composition, a sieve made of hair is generally used, and for less dangerous compositions a sieve made of copper wires may be used. Apertures of 800–1000$\mu$ are usual for mixing sieves. The ingredients which cohere easily, (for example potassium chlorate, barium nitrate, potassium nitrate, colophony) should be sieved just before the mixing process. (Potassium chlorate can now be purchased as a free flowing powder).

As an example we can cite here the mixing process of composition H3, the name of which comes from the content of hemp coal it contains (i.e. the ratio of hemp coal to potassium chlorate is 3:10). H3 is sometimes used as a bursting charge for shells in Japan. This process applies also in the same way to other less dangerous compositions. A kraft paper is laid on a work table. (A sheet of synthetic resin is not recommended for this purpose, because it is too soft to handle). The paper should be used only once and not be used for other compositions, so it would be best to use a kind of cheap cement bag paper. The prescribed amount of potassium chlorate, hemp coal and glutinous rice powder are weighed on to the paper. The ratio of the three components is generally 10:3:0.3. A part of the hemp coal and all the rice powder are added to the potassium chlorate and mixed well by hand, and then it is passed through a hair sieve once. The potassium chlorate then becomes free flowing and can be mixed easily. The residue in the sieve should be left untouched, and not rubbed with the hands. The rest of the hemp coal is all mixed, at one time, into the premixed potassium chlorate and allowed to pass through the hair sieve two or three times. In this case the residue in the sieve is also left untouched. The last residue which remains in the sieve after the mixing process is thrown away in a suitable place of safety. The finished composition is packed into a prescribed container, which bears a label securely affixed thereto, showing the kind of composition, the formula, the date of manufacture, the name of the worker, and is carried into an interim store house. It is forbidden to leave such an explosive or inflammable product in the work room. The quantity of material which can be mixed at one time depends upon the composition. In the case of sensitive compositions it should be less than 5 kg. To mix more than this amount by hand perfectly takes a longer time, and causes more danger. In place of hand mixing it is possible to use a V type or cylinder type mixer, but special care must be taken in charging or discharging its content.

In the case of black powder, even when the three components (potassium nitrate, sulphur and charcoal) are mixed together by the above process, the force of the powder is very small. In the case of black powder production as an industrial explosive on a large scale, the three components are brought together by pressing or milling. Beating in a mortar to manufacture black powder in small amounts for fireworks is very dangerous, and the semi-wet process is much easier to handle. 7.5kg. of potassium nitrate and 2.5kg. of water are charged into an iron pan and heated to 100°C, stirring constantly. Hot water may also be used to heat the pan. Next the contents are taken out of the pan and charged into a mixer. 1.5kg. of hemp charcoal is then added and the two are mixed for an hour to soak the hemp coal in the potassium nitrate. This process is safe, because there is still some moisture in the mixture. Then the mixture is dried and charged into a ball mill, which is made of wood in an eight angle drum of 33 cm. in diameter and 35 cm. in length. This is placed in a work room surrounded by defense walls. 1.0kg. of sulphur is added to the mixture and then about 100 glassy porcelain balls, each of which is 33mm. in diameter and 30 grams in weight; these are charged into the drum which is driven for 24 hours. Such a three component mixing process serves two purposes, i.e. mixing and milling. It is a dangerous operation and not always recommendable, but it has been used for a long time in my factory without accident. The "three component" mixing room should be capable of bursting in one direction and the operator should not enter the room while the drum is revolving. The force of the black powder produced by the above process was compared with that of ordinary hand-mixed black powder by comparing their burning ratios in the atmosphere. If we take the value of the hand mixed powder as 1, then the value of the powder made by the above method in the wet state becomes 12, and "three component" dry mixing gives a value of 6. This proves the superiority of the semi-wet process described above. Never use a porcelain mortar or drum for mixing the three components, because it is very dangerous.

*Bursting Charge*

The bursting charge is the powder which is charged into a shell and which breaks it by means of an explosion. The purpose is to throw out end spread the contents of the shell, to ignite the contents and to give velocity to the stars. We have two kind of shells, the so-called "Warimono" and "Poka". Warimono is a chrysanthemum shell which needs a fairly large amount of strong bursting charge. It is capable of giving proper velocity to stars so that they form a round flower in the sky. Poka is an ordinary shell like a flag, willow or flare which does not need so much or so strong a charge as Warimono.

The force of the charge for Poka should be enough only to break the shell, and the black powder on the market or powder made in the factory is generally used for the purpose. The powder is mixed with water to a paste and chaff is covered with the slurry.

The bursting charge for Warimono should have a large explosive force and proper burning rate. When the value for the burning rate is too large, the stars in the shell are destroyed or not ignited. On the contrary when the value of the burning rate is too low we cannot have a good radius of fire in the chrysanthemum.

*Potassium Perchlorate Bursting Charge I*

|                                | %                      |
| ------------------------------ | ---------------------- |
| Potassium Perchlorate          | 70                     |
| Hemp Coal                      | 18                     |
| Sulphur                        | 12                     |
| Soluble Glutinous Rice Starch  | 2 (Additional Percent) |

This formula has already been used for 15 years and no accident has happened. It is applicable for practical use. It is thought that it is safer to handle than the potassium chlorate bursting charge. A 2 kilogram drop hammer test showed that the non explosive point is 37 cm. (In the case of picric acid : 23 cm.) By Yamada's friction test apparatus, we obtained the non explosive weight of 75 kg. (In the case of picric acid : 56 kg.) The burning rate of this composition in the atmosphere when it is pasted on cotton seeds is 1.0 mm/sec. and is rather small. A counted value of the force of explosion shows $0.71 \times 10^6$ dm. This value is twice that of black powder. The values of the heat of explosion, the specific volume (= the volume of the gas created by explosion measured at 0 °C, 1 atm.) and the explosion temperature are 690 kcal/kg, 480 l/kg and 3000 °C respectively. Recently the author has proposed the use of the following formula, which is thought to be safer than the former because of the absence of sulphur.

*Potassium Perchlorate Bursting Charge II*

|                                | %                      |
| ------------------------------ | ---------------------- |
| Potassium Perchlorate          | 70                     |
| Hemp Coal                      | 25                     |
| Lamp Black (Pine Black)        | 5                      |
| Potassium Bichromate           | 5 (Additional Percent) |
| Soluble Glutinous Rice Starch  | 2 (Additional Percent) |

Generally it is better to use very fine charcoal to get a high burning rate, i.e. the finer the better. The force of explosion of the above formula seems to be somewhat smaller than that of the former. The values of the non-explosive point and the non-explosive weight, by the same test apparatus, are 60 cm. and 75 kg. respectively.

*Potassium Chlorate Bursting Charge H3*

|                                | %                      |
| ------------------------------ | ---------------------- |
| Potassium Chlorate             | 77                     |
| Hemp Coal                      | 23                     |
| Soluble Glutinous Rice Starch  | 2 (Additional percent) |

This is the famous old formula of the bursting charge H3, which is still widely used at present in Japan. The ratio of hemp coal to potassium chlorate is about 3:10, from which the name H3 comes. Compositions, which come into contact with this formula, increase its sensitivity when they contain sulphur, realgar, red phosphorus or antimony trisulphide, and we must pay close attention to this in production management. Compositions of this kind are thought to be the ones which most frequently cause accidents. Many fireworkers replace some part of the potassium chlorate with potassium nitrate, but it is not always certain to be safe enough. The value of the non-explosive point obtained by a 2 kilogram drop hammer test was 50 cm. By Yamada's friction test apparatus we had the value of the non-explosive weight of 75 kg. The latter is the same as that of the potassium perchlorate charge. The value of the burning rate of this composition pasted on cotton seeds was 3 mm/sec. at 1 atm. This value is three times as large as that of the potassium perchlorate composition. The counted values of the force of explosion, the heat of explosion, the specific volume and explosion temperature are $0.74 \times 10^6$ dm, 640 kcal/kg, 570 l/kg and 3100°C respectively. This composition has been widely used because of its large explosive force and the simplicity of the manufacturing process. One must not mix it with black powder, but both compositions have the same black color so care must be taken not to take the wrong one.

### Potassium Nitrate Bursting Charge

The formula is no different from the normal formula for black powder (i.e. the weight ratio of potassium nitrate, charcoal and sulphur is 75:15:10). As a pasting material 2% of glutinous rice starch is added additionally. It is thought to be the safest composition with regard to friction and shock. The 2 kilogram drop hammer test showed the non-explosion point of 85 cm. The burning rate of grains made of cotton seeds, pasted with this composition, is about 2 mm/sec. at 1 atm. This value is twice as large as that of the potassium perchlorate composition—in practice, in place of charcoal we use hemp coal. It contains 10% moisture and 10% ash, and the pure coal remains as 80%. The counted value of the force of explosion is $0.3 \times 10^6$ dm, which is smaller than ordinary rifle powder. This is because a smaller volume of explosive gas is created. The values of the heat of explosion, the specific volume of the explosion gas and the explosion temperature are counted as 400 kcal/kg, 370 l/kg. and 1800 °C respectively. This composition is extremely ignitable and has a large value for the burning rate, but its disadvantage is that the force of explosion is too small It. is used for relatively large shells.

To make the fire spread rapidly when the shell is ignited, the composition is pasted on to a nucleus. Cotton seeds, chaff or immature ears of rice are most frequently used as nuclei. In the case of cotton seeds the ratio of the composition is 1:1.3—1.6 and in the case of chaff its ratio to the composition increases. In general the charge made of cotton seeds is used for relatively large shells which need only a small loading density, and that made of chaff is used for small shells which need a relatively high loading density. Sometimes these are mixed together. For example:

*The Grains of Bursting Charge*

For Small Shells (less than 3 1/2 inches in diameter)
    Composition                        80%
    Rice Chaff                         20%

For Intermediate Shells (from 5 to 7 inches)
    Composition (Potassium Perchlorate Bursting Charge) 52%
    Cotton Seeds                       48%

For Large Shells (from 8 to 12 inches in diameter)
    Composition (Potassium Perchlorate or Nitrate
                Bursting Charge)         52%
    Cotton Seeds                       48%

The composition is poured, with the nuclei and a suitable amount of water, into a vessel and stirred until the nuclei are covered with the composition. The grains produced in this manner are dried in the sun. The thickness of the pasted composition on the nuclei is irregular in the case of chaff but is 0.47—0.49 mm. on average in the case of cotton seeds. More than this is not recommended because of its performance character as the bursting charge. Bursting charges which follow the above formulae show an apparent specific gravity of 0.48—0.50. The loading density of the bursting charge is counted, rejecting the nuclei in it, which have values of 0.38—0.40 and 0.25—0.26, in the case of cotton seeds and chaff respectively. For example the amounts of the bursting charge used in shells, without nuclei according to the above specification are follows:

*The Amounts of Bursting Charge for Chrysanthemum Shells*

| Diameter of the Shell inches. | Amount of the Bursting Charge grams. |
|---|---|
| 3 1/2 | 44 |
| 5 | 70 |
| 6 | 140 |
| 7 | 290 |
| 8 1/2 | 400 |
| 9 1/2 | 940 |
| 12 | 1250 |

When cotton seeds and chaff are used as the nuclei they should be well dried beforehand.

*Colored flame compositions*

In general a colored flame composition consists of oxidizer, fuel, and color-creating material. Some metal salts play the role of both oxidizer and color-creating material (e.g. barium nitrate, barium chlorate and strontium nitrate).

*Oxidizer*

When ammonium perchlorate is used as an oxidizer, it produces a relatively small amount of smoke in air of low moisture content, because it creates very little solid material during burning. In moist air the burning gas produces small water particles and it creates a dense smoke cloud. Ammonium perchlorate gives the most clear and beautiful coloured flame, when it is used with all the color-creating materials. For frame fireworks, ammonium perchlorate composition is most recommended if the cost will allow this. The contact of ammonium perchlorate composition with black powder or potassium nitrate creates ammonium nitrate, which is very hygroscopic and damages the fireworks. Ammonium perchlorate composition is very ignitable and generally requires no igniting composition.

Potassium perchlorate composition creates more smoke and less depth of color than ammonium perchlorate composition, but is more practicable. The fuel which is used to the best advantage in ammonium perchlorate compositions will not always give the same effect in potassium perchlorate compositions. When potassium chlorate is used as an oxidizer, the effect is not so different as that of potassium perchlorate, but the amount of smoke seems to be larger than the former and, around the base of the flame, potassium chlorate creates a tubular ash, which disturbs the flame projection. For example in the case of a lance, the flame does not come out of the end of the lance but branches out as it is diverted by the ash. The tubular ash also diminishes the light intensity of the flame.

The color of the flame depends upon the kind of oxidizer, even when we use the same fuel and color-creating material. For example an ammonium perchlorate composition with strontium carbonate as the color-creating material can give a very clear and deep flame, but in the case of potassium perchlorate in place of ammonium perchlorate the flame looks reddish pink.

As for the fuel, it is advisable to use a material which gives no color to the flame, but such material cannot be obtained in practice, and we must be content with a material which gives as weak a color to the flame as possible.

Using ammonium perchlorate as the oxidizer, fuels were tested to discover which were least likely to disturb the flame color. The order of preference was as follows: Shellac, wood meal, pine root pitch, amber powder, colophony, charcoal. Samples, 115 mm. in length and 7 mm. in diameter, each of which consists of a brown paper tube and 6 grams of a composition pressed in it to an apparent specific gravity of about 1.35, are ignited one by one at one end and their burning state is observed. The composition consists of 17% of a fuel and 83% of ammonium perchlorate, and contains no color-creating material so that the color disturbing effect of the fuels can be observed.

The result is summarized as follows:

| Kind of Fuel | Flame Condition | Burning Rate | Flame Length |
|---|---|---|---|
| Shellac | Weak reddish orange | 1.51 mm/sec. | 90 mm. |
| Wood Meal | Weak reddish orange lines | 1.27 " | 60 mm. |
| Pine Root Pitch | Bright white | 2.09 " | 110 mm. |
| Amber Powder | Brilliant white at base | 1.44 " | 70 mm. |
| Colophony | Brilliant showing lines | 1.46 " | 100 mm. |
| Hemp Coal | Yellow and brilliant | 2.80 " | 90 mm. |

For potassium perchlorate the most suitable fuel is pine root pitch, and then it should be arranged in order of merit as follows: Colophony, amber powder, hemp coal, wood meal. Shellac and potassium perchlorate burn with some difficulty. In the same manner as described above, with a composition of 17% of a fuel and 73% of potassium perchlorate, we obtained the result as follows:

| Kind of Fuel | Flame Condition | Burning Rate | Flame Length |
|---|---|---|---|
| Pine Root Pitch | Weak violet at the top and brilliant at the base | 1.83 mm/sec. | 80 mm. |
| Colophony | Weak violet at the top and the rest white | 1.24 '' | 50 mm. |
| Amber Powder | Accompanied by much white smoke | 1.51 '' | 60 mm. |
| Hemp Coal | Yellow at the center and the rest reddish | 4.80 '' | — |
| Wood Meal | Violet white, burning with difficulty | 0.69 '' | 20 mm. |
| Shellac | White, burning with difficulty. | 0.84 '' | 30 mm. |

These compositions create much smoke. When we use wood meal or shellac with potassium perchlorate, we must also add other more combustible fuels.

When potassium chlorate is the oxidizer, pine root pitch is the most suitable fuel and shellac follows next. Colophony burns with difficulty. Amber powder gives a white flame and much smoke. The mixture of charcoal and potassium chlorate burns very rapidly and dangerously, and charcoal alone is seldom used as the fuel. Charcoal is generally used to adjust the burning rate of composition, to deepen the color of the flame, to increase its brilliancy or to be a supplementary fuel for some other purpose. The result of a test for compositions, in which 83% of potassium chlorate and 17% of a fuel are used in the same way as described above, is as follows:

| Kind of Fuel | Flame Condition | Burning Rate | Flame Length |
|---|---|---|---|
| Pine Root Pitch | Much smoke and violet flame | 2.13 mm/sec. | — |
| Shellac | Light violet, ash remains | 1.20 '' | 80 mm. |
| Amber Powder | Much smoke, white flame | 1.60 '' | — |
| Colophony | No burning | 2.26 '' | — |
| Wood Meal | Violet flame, vibrational burning | 2.26 '' | 30 mm. |
| | — | 17.4 '' | — |

*Hemp Coal (Charcoal)*

Coal tar pitch disturbs the color of the flame remarkably and cannot be used as a fuel for colored flame compositions.

*Color*

Color creating materials, have the following characteristics:

Red color creating materials: Strontium carbonate is one of the red color creating materials, and it gives the most beautiful red flame, but a high percentage in a composition causes the generation of very much smoke and creates burning difficulties. The effect of strontium oxalate is the same as strontium carbonate. Strontium nitrate gives a very beautiful red flame but it is a weakness of strontium nitrate that it is rather hygroscopic.

Yellow color creating materials: Sodium oxalate gives a clear yellow color to the flame even in small amounts, but the color is somewhat reddish and not pure yellow. Borax gives almost pure yellow. Ultramarine leaves a residue on burning, and is not so commendable.

Green color creating materials: Barium chlorate gives the deepest green, but creates much smoke when the amount is increased. This material resolves easily, and is not recommended. Barium nitrate cannot give the depth of color, but when we have a good formula, it can give a green flame of practical use.

Blue color creating materials: Copper sulphate gives a good blue color to the flame in an ammonium perchlorate composition, and its effect is not inferior to that of emerald green (Paris green). Copper sulphate powder is neither as fine nor as good as that of Paris green. As a safety precaution it is forbidden to mix copper sulphate with potassium chlorate or other chlorates. Paris green will give a good flame color with any kind of oxidizer. The only defect is that Paris green is easily scattered in the air, and it is poisonous when the worker inhales it. Copper arsenite shows almost the same character as Paris green in handling and in coloring the flame.

As an ingredient of low cost, calcium carbonate is sometimes used in place of other red color creating materials. In this case the flame is reddish orange and not so beautiful.

The colored flames for fireworks are founded on the useful band or line spectra, which are inherent in metal or metal compounds. But the color condition, which appears to the eyes, depends upon not only the strength of the spectra caused by the color creating material, but also the following conditions: Every flame has background spectra, which depend upon the kinds of metal salts and fuels used in the composition.

The flame temperature has also an influence upon the flame color. The paper tube, which contains the composition sometimes disturbs the coloring of the flames. The physiological condition of eyes, moisture, dust or mist in the air change the color of the flame. Note that an electric light which is white at a close distance looks reddish yellow at a distance of 1—2 kilometers. A green star, which looks light yellowish green at a close distance and appears to be useless, gives a deep and clear, beautiful green when we observe it at a distance of 50 meters. It is one of the general rules when designing the formula of a star composition that we must observe the burning star at a long distance.

The principle of designing colored flame compositions is as follows: First the oxidizer is selected according to the use (e.g. for stars, frame fireworks, flares) and then the color creating material and fuel which are suitable for the oxidizer. The flame color should be carefully chosen so that it is the best for the purpose. The fuel should be carefully arranged to give the most adequate burning rate. To get a light of high intensity the flame temperature must be as high as possible, and the composition should be rich in oxygen content or should contain a metal fuel like magnesium powder. Chlorine gas in the flame deepens the color of the flame, especially when barium and copper salts (in the case of green or blue flame) are used as the color creating material. Potassium nitrate is not used generally as the oxidizer in colored flame composition, because it gives a low burning temperature which is insufficient to excite the coloring molecules or atoms in the flame. Ammonium perchlorate is not used in practice for shell stars because of its hygroscopic nature and difficulties in ignition at high velocity in the air. In the case of shell stars the composition may be allowed to create smoke to some extent and the flame is somewhat inferior in color, but should be well dried and not be hygroscopic. The compositions for frame fireworks should not create so much smoke and their tendency to water absorption to some degree may not disturb their ignitability because they are not exposed to strong winds as in the case of shells. Moreover they are loaded as powder, which is very ignitable.

When we solidify and form the composition into a certain shape for use, we must add some adequate solidifying material, a so-called "binder" to the composition. When water is used as a solvent, soluble glutinous rice starch is recommended, and shellac is used when alcohol is the solvent. If water is forbidden it may be possible to use linseed oil or other drying oil, and in this case care must be taken to disperse the heat of oxidation of the oil. Celluloid dissolved in amyl acetate can also be used as a binder. The amount of these binders should be as small as possible.

For example colored flame compositions of practical use are shown as follows:

*Colored Flame Compositions for Frame Fireworks (lances)*

| | % |
|---|---|
| Red Flame: | |
| Ammonium Perchlorate | 70 |
| Strontium Carbonate | 10 |
| Wood Meal | 20 |
| Yellow Flame: | |
| Ammonium Perchlorate | 75 |
| Sodium Oxalate | 5 |
| Wood Meal | 15 |
| Colophony | 5 |
| Green Flame: | |
| Ammonium Perchlorate | 50 |
| Barium Nitrate | 34 |
| Wood Meal | 8 |
| Shellac | 8 |
| Blue Flame: | |
| Ammonium Perchlorate | 70 |
| Copper Sulphate | 10 |
| Wood Meal | 10 |
| Shellac | 10 |
| White Flame: | |
| Ammonium Perchlorate | 40 |
| Potassium Perchlorate | 30 |
| Antimony Trisulphide | 14 |
| Starch | 11 |
| Wood Meal | 5 |

Each of these compositions is loaded in a thin brown paper tube of 9mm in diameter with a density of 1.3—1.4 and so designed that the rate of burning should be 1.0—1.1 mm/sec. Generally when a powdered composition is pressed by hand into a paper tube and ignited at one end, the relation between the burning rate and the diameter of the tube differs with the type of composition, and the greater the diameter, the greater the burning rate below a diameter of 20 mm. For larger sizes than this limit, the burning surface becomes unsteady and causes flake dropping, When we take the burning rate at the diameter of 10

mm. as the unit 1, the burning rate at the diameter of 20 mm. reaches
1.7—3.0.

*Color Flame Compositions for Stars*

Red Flame:

| | |
|---|---|
| Potassium Perchlorate | 67 |
| Pine Root Pitch | 13.5 |
| Strontium Carbonate | 13.5 |
| Soluble Glutinous Rice Starch | 6 |

Yellow Flame:

| | |
|---|---|
| Potassium Perchlorate | 72 |
| Sodium Oxalate | 7 |
| Pine Root Pitch | 12 |
| Colophony | 3 |
| Soluble Glutinous Rice Starch | 6 |

Green Flame:

| | |
|---|---|
| Potassium Perchlorate | 46 |
| Barium Nitrate | 32 |
| Pine Root Pitch | 16 |
| Soluble Glutinous Rice Starch | 6 |

Blue Flame:

| | |
|---|---|
| Potassium Perchlorate | 64 |
| Pine Root Pitch | 13 |
| Paris Green | 17 |
| Soluble Glutinous Rice Starch | 6 |

In the case of stars for chrysanthemum shells the designing conditions depend especially upon the burning rate of the stars. Pine root
pitch is used here to make the burning as fast as possible, and if there
is another fuel which gives the same character, it is used in place of
pine root pitch because the pitch is difficult to get. In order to obtain
a good chrysanthemum, fuels must be chosen so that the compositions
burn quickly without disturbing the flame color. The color of the green
flame is not so good at a short distance, but is practicable from a long
distance according to the physiological phenomena of the eyes. If the
relation of the burning rates between the solidified composition, which
is made by the wet method, and that of pressed powder composition
of the same formula were previously studied, it would be better to
design the stars from the data of the pressed composition, which is
easier to make. Insufficient study has been made hitherto, but a few
examples are shown in Table A.

*Table A.* Comparison between the burning rates of solidified and pressed compositions.

|  | Solidified Star | | Pressed Star | |
|---|---|---|---|---|
|  | S.G. | B.R. | L.D. | B.R. |
| A Blue Star | 1.60 | 1.79 mm/sec. | 1.47 | 1.70 mm/sec. |
| A Green Star | 1.64 | 1.18      ” | 1.60 | 1.52      ” |

(Here S.G. = Specific gravity;  B.R. = burning rate;
   L.D. = Loading density)

Each of the solidified stars is a sphere of 12 mm. in diameter, and the pressed stars are cylindrical using a paper tube of 9 mm. in diameter, into which the powder is pressed. The table shows that the relation between the burning rates of the solid and powdered state depends upon the kind of composition, and in practice the influence of the solidifying material (binder) must be considered. Studies of the burning rates of stars of different diameter but of the same composition have not yet been sufficiently made, but we know that the larger the diameter is, the larger is the burning rate. A few examples are shown in Table B.

*Table B.* The relation between the diameter and burning rate of spherical and cylindrical stars.

| *Spherical Stars* (solidified) | | |
|---|---|---|
| Diameter | 18 mm. | 32 mm. |
| A Blue Star | 1.79 mm/sec. | 1.92 mm/sec. |
| Some Star | 3.89      ” |  |

| *Cylindrical Stars* (pressed) | | |
|---|---|---|
| Diameter | 9 mm. | 21 mm. |
| White Star | 1.95 mm/sec. | 2.10 mm/sec. |
| Yellow Star | 1.15      ” | 3.00      ” |
| Red Star | 0.98      ” | 1.80      ” |
| Blue Star | 1.00      ” | 1.28      ” |
| Green Star | 1.04      ” | 1.60      ” |

In the case of the composition for shell stars (e.g. for color changing stars) there are many successive ignitions from one composition to another. For ignition without failure the amount of heat of burning, and the burning rate of the former layer and the specific heat of the latter layer, must be adequate. The star compositions for shells described above do not have very good ignition because of the potassium perchlorate, and the construction of stars requires some device to make them ignite easily.

Generally each of the colored flame compositions has a sufficient quantity of oxygen to oxidize the fuel it contains, and it burns progressively in the solidified or pressed state when it is ignited, but causes detonation when it is initiated with a percussion cap. We must pay close attention to this characteristic of color composition.

*Fire dust spark composition*

The colored flame is admired from both short and long distances, but the sparks described here, which are like fire dust, are admired at a long distance. The fire dust sparks are the phenomena caused by charcoal, metal and some ash which are projected at high temperature from a burning composition as unreacted matter, or newly formed matter, and make a second burning with oxygen in the air. This can be verified by burning the fire dust over a vessel of water when it will be possible to detect the unreacted or yet reactable charcoal, metal or ash in the water.

The composition for fire dust sparks must not contain enough oxygen to oxidize the fuel it contains. (Viz. the composition for fire dust sparks consists of a smaller amount of oxidizer and a larger amount of fuel than colored flame composition.)

The colored flame is admired for its spectra which is peculiar to the metal or the metal compound in the vapor phase. But in the case of fire dust sparks it is concerned with the temperature of solid or liquid particles. The color of the light emitted by a radiator changes according to its temperature, and in ordinary cases it follows thus:

| | |
|---|---|
| 550 °C | dark red |
| 700 °C | red |
| 850 °C | orange |
| 1000 °C | yellow |
| 1100 °C | white |

To obtain beautiful fire dust sparks it is necessary to adjust the combustion temperature of the composition.

Other problems arise, though, with fire dust stars. Are both the ignition characteristics and the separation of the fire, good or not? And generally, as regards appearances, the particle size of the fire dust, density of the fire particles and the lifetime of the particles are the most important factors. When planning fire dust spark compositions we must have a clear knowledge of these conditions, but this has not yet been studied fully and systematically, and here we can describe only a few fragments, which have been made clear. In compositions of this kind we often see somewhat complicated formulas, but is seems that they depend upon the conditions described above and especially upon the regulation of the burning temperature of the compositions. Now we shall study the influence of the components upon the temperature of some compositions of the simplest formula.

The character of the fire dust sparks caused by charcoal depends upon the formula, the character of the charcoal and its particle size. It is impossible to adjust the color of the fire dust very much, for the color at best only lies between reddish orange and orange red. The composition of potassium nitrate, sulphur and charcoal, can produce the most beautiful fire dust sparks, but composition which does not contain potassium nitrate cannot succeed in producing such sparks. Charcoal, for example, with ammonium perchlorate, burns perfectly in the flame without discharging sparks. In the case of potassium chlorate it is the same as above, except that the burning reaction is more violent than the former. With potassium perchlorate it creates a few sparks, but most of the reaction finishes in the flame. Only in the case of potassium nitrate, where the flame is relatively small, can a large amount of fire dust sparks be produced out of the flame. It seems that the potassium sulphide produced in the flame surrounds the particles of charcoal and disturbs the strong reaction in the flame. The smaller the particle size of the charcoal, the shorter will be the life of the fire dust sparks, and the phenomenon also depends upon the amount of sulphur and the kind of charcoal contained in the composition. Some examples of the formula are as follows. (These compositions are called "Tail" or "Chrysanthemum".)

*Chrysanthemum 6.*

|  | % |
|---|---|
| Potassium Nitrate | 58 |
| Sulphur | 7 |
| Pine Charcoal | 35 |

*Chrysanthemum 8.*

|  |  |
|---|---|
| Potassium Nitrate | 52 |
| Sulphur | 6 |
| Pine Charcoal | 42 |

The number six or eight means that the weight of pine charcoal is 6 or

8 relative to 10 parts of potassium nitrate. When a high velocity of burning or a large quantity of heat is required, chrysanthemum 6 is used, and for the opposite requirement chrysanthemum 8 is used. An intermediate composition or compositions outside this range are also used. To solidify the stars soluble glutinous rice starch is added as a binder (The percentage is normally about 6%). To make a composition of very fine fire dust, which can produce sparks of high concentration, the above formula is prepared and charged into a wooden ball mill with porcelain balls and it is then driven for 24 hours. The life of the sparks is short in this case but they are very beautiful and elegant. When the composition has no sulphur, it is difficult to burn well, and the fire dust branching out is not so good, for the stars keep the fire themselves for a long time. This phenomenon produced by moving stars looks like the dawning of a weak fire dust band or tail in the sky, and it appears peculiarly elegant. At a short distance it is observed that many fragments of fire are falling down. This kind of star is called "Chrysanthemum of Mystery".

| Chrysanthemum of Mystery. | % |
|---|---|
| Pine Charcoal | 50 |
| Potassium Nitrate | 45 |
| Soluble Glutinous Rice Starch | 5 |

Sometimes 3—4% of minium is added to this composition. The purpose may be to use the special after reaction which is peculiar to minium, or to increase the weight of stars, but it seems that adding minium has little influence upon the fire dust sparks phenomenon. Other kinds of charcoal and other formulas are also used. The composition which contains lamp black produces sparks with a reddish orange flame.

The defect of the composition made of potassium nitrate, sulphur and charcoal is that it has less ability to ignite the next layer of the star than that of the colored flame compositions, because of the small calorific value and low burning temperature. This would be expected because of the lack of oxygen in the composition. Therefore when it is hard to ignite the second layer from the first layer, (which consists of a black powder type composition), it is necessary to put a new layer of a more easily ignitable and hot composition between the first and the second layers.

As a source of metal fire dust sparks we usually use aluminum powder, and mix it with potassium chlorate, potassium perchlorate, potassium nitrate or barium nitrate, but from the stand point of ignition and safety, chlorate is not recommended. The sparks from these sources have a peculiar beauty. Sulphur, realgar, antimony trisulphide, charcoal are added to regulate the temperature of the sparks or to give good ignition to the stars.

Potassium nitrate and aluminum spark composition produce beautiful fire dust sparks which are reddish orange, gold or yellow white.

The so called "Fire Dust Branching" phenomenon occurs when a melted mass of high temperature branches out into fine fire dust by the action of air currents. To see this phenomenon it is important to test the burning star by projecting it in the sky. The star, which burns on the ground and generates very good sparks, does not give such a good effect in the sky, because the life of the sparks looks too short. In the case of the ground test it is seen that the melted ash after the burning reaction becomes a mass and has brightness without producing many sparks, though it creates good sparks in the air. On the other hand it should be noted that the overmelted ash does not branch into much fire dust; in fact the life of the sparks becomes too long and the beauty diminishes. It is the same when potassium perchlorate is the oxidizer.

*Falls:*

|                              | %  |
|------------------------------|----|
| Potassium Nitrate            | 41 |
| Aluminum                     | 49 |
| Sulphur                      | 4  |
| Soluble Glutinous Rice Starch | 6  |

A small amount of water is added to the composition and it is kneaded well; it is then pressed into a thin paper tube with a wooden or brass pounder. It is said generally that thorough kneading gives a long life to the sparks, but it does not always seem to be true. Coarse aluminum flakes are good for making the sparks burn long enough, and sometimes we use aluminum foil flakes (flitter) in the falls composition. When the tube has been loaded with the composition it is dried in the sun. When it is noticed that the tubes are generating heat, they must be separated from each other in the shade to cool them. If a long time is taken to dry the pieces, the fire dust sparks which may be produced will become somewhat reddish. The burning rate of a waterfall of 15 mm. in diameter was 0.52 mm/sec. This falls composition can produce fire dust sparks of silver color which reach to the ground, hanging at a height of 3 meters. It is used for set fireworks "Falls".

Composition of this kind, which contains nitrate and aluminum and a small amount of water sometimes generates enough heat to catch fire, when it is left for a long time in the wet state. The heat comes from the reaction between nitrate and aluminum, which creates $H_2$, $NH_3$, NO or $NO_2$. To avoid accidents never leave the composition in the wet state.

Generally it is difficult to ignite stars, which consist of the two components, i.e. potassium nitrate and aluminum. When the amount of aluminum increases to 50%, it is impossible to ignite it by ordinary methods, but on the other hand the larger the amount of aluminum, the more beautiful the sparks are, and we add other ingredients to the composition to get good ignition, even if it contains a large amount of aluminum. Sulphur is the most suitable ignition ingredient, but it seems to create trouble when stored for long periods. Next, in order of preference, antimony trisulphide or charcoal is recommended, but realgar is not so good for this purpose. The color of the fire dust sparks depends upon the burning temperature, and accordingly it is peculiar to the kind of fuel used. The following compositions are all of practical use, but note that they are difficult to ignite when they are made into stars, and it is necessary to cover the stars with an easily ignitable composition which is specially selected.

*Golden Wave No. 1*

|                               | %  |                        |
| ----------------------------- | -- | ---------------------- |
| Potassium Nitrate             | 40 |                        |
| Aluminium                     | 50 |                        |
| Antimony Trisulphide          | 10 |                        |
| Soluble Glutinous Rice Starch | 5  | (Additional percent)   |

*Golden Wave No. 2*

|                               | %  |                        |
| ----------------------------- | -- | ---------------------- |
| Potassium Nitrate             | 40 |                        |
| Aluminum                      | 50 |                        |
| Sulphur                       | 10 |                        |
| Soluble Glutinous Rice Starch | 5  | (Additional percent)   |

*Golden Wave No. 3*

|                               | %  |                        |
| ----------------------------- | -- | ---------------------- |
| Potassium Nitrate             | 40 |                        |
| Aluminum                      | 50 |                        |
| Realgar                       | 10 |                        |
| Soluble Glutinous Rice Starch | 5  | (Additional percent)   |

Compositions, No. 1 and No. 2. produce very fine golden fire dust sparks. Composition No. 3 produces reddish golden sparks, which are of special elegance and different from the sparks of No. 1 and No. 2. Potassium perchlorate and aluminum spark composition: The composition of potassium perchlorate and aluminum is more ignitable than the above potassium nitrate compositions, and it can be used without adding other fuel. We can increase the amount of aluminum to 75% without losing its ignitability, but when the amount of aluminum is over 65%, the fire dust branching is not so good. The practical compositions are as follows:

*Silver Waves:*

|  | No.1 % | No.2 % | No.3 % | No.4 % |
|---|---|---|---|---|
| Potassium Perchlorate | 50 | 45 | 40 | 35 |
| Aluminum | 50 | 55 | 60 | 65 |
| Soluble Glutinous Rice Starch | 5 | 5 | 5 | 5 |
|  |  |  |  | (Additional percent) |

The amount of soluble glutinous rice starch in the formulas above is also suitable for pressed stars, and in round stars it should be 8—10% for ease of manufacture, but in this instance they are more difficult to ignite. The compositions which contain metal powder like golden waves or silver waves are less easily ignited than ordinary compositions for stars, and must be formed by means of a special process described later. (See Page 249). Composition No. 1 appears with a short silver stream which has a short life in the air. As the composition changes its constitution from No.2. to No.4. the brilliancy of the stars gradually diminishes and the stream of the sparks becomes larger and longer.

Magnesium cannot produce such good fire dust sparks as aluminum, when it is used with potassium nitrate or potassium perchlorate, because magnesium is vaporized at the relatively low temperature of about 1100 °C and burns completely without producing hot liquid or solid matter which can be projected out of the flame.

The phenomenon of the fire dust sparks is somewhat complicated, but by the application of the fundamental phenomenon described above it may be possible to plan various compositions of practical use.

*Fire-branching sparks composition*
The so-called "Fire-Branching Sparks" is a phenomenon produced when a liquid or solid matter of high temperature is projected into the air and explosively separates into many fire branches like a pine needle.

It is therefore different from the fire dust sparks described above, which do not branch explosively. This kind of spark is seen at short distances. There are two kinds of composition for the sparks: one contains charcoal and the other contains iron or magnesium powder. The former belongs to the toy fireworks "Senko Hanabi" and the latter to "Fountains" and "Sparklers" etc.

The spark composition using charcoal consists of potassium nitrate, sulphur and charcoal and belongs to the black powder type of composition, but the peculiar distinction from the black powder is that the amount of charcoal is less than the amount of sulphur. The composition which produces good sparks contains 10—15% of charcoal. It seems that any kind of charcoal or soot, except crystallized carbon like graphite, can produce the pine needle-like sparks, but the finer the particles of carbon, the better the sparks. Accordingly pine soot is the best, followed by paulownia charcoal or hemp coal. Pine charcoal can also produce the sparks, which are not so large as those of pine soot. A test showed that carbon black produces almost no sparks, though it does create, on rare occasions, very large ones. The examples of the formulas used in old times are shown as follows:

*Senko Hanabi (Japanese Sparkler):*

|                     | No.1 % | No.2 % | No.3 % | No.4 % | No.5 % |
| ------------------- | ------ | ------ | ------ | ------ | ------ |
| Potassium Nitrate   | 50     | 61     | 36     | 48     | 59     |
| Sulphur             | 34     | 24     | 22     | 34     | 24     |
| Pine Soot           | —      | 15     | 7      | 12     | 14     |
| Paulownia Charcoal  | 8      | —      | 35     | —      | 3      |
| Pine Charcoal       | 8      | —      | —      | 6      | —      |

In the author's laboratory the compositions were studied, by observing the sparks which are projected from hot liquid drops produced by burning compositions, (of pine soot, potassium nitrate and sulphur) which are pressed in thin paper tubes. The results are shown with a triangular graph as in Fig. 113. The area with single shadow shows the spark producing zone while the double shadow shows the zone with excellent sparks.

When we use other oxidizers in place of potassium nitrate, we cannot find such pine needle-like sparks, which are peculiar to Senko Hanabi. Sulphur is replaceable by realgar, but other materials are unable to produce such sparks. Charcoal is not replaceable by other materials.

Potassium
Nitrate

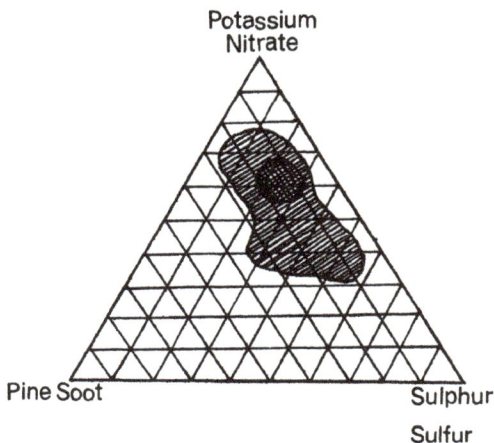

Pine Soot                                    Sulphur

Sulfur

**Fig. 113**   Japanese sparkler compositions

To manufacture Senko Hanabi (Japanese Sparkler) about 0.1 gram of the composition is twisted in a Japanese paper tape (20 cms. × 2.5 cms) or conglutinated on one end of a rush halm. When it is ignited, it burns violently with a flame at first; then the remaining ash keeping its red-hot state, contracts itself to a red-hot small ball, which is the so-called "Fire Ball". After a few seconds the temperature of the fire-ball gradually rises and fine particles begin to fly out of the ball. The particles become more brilliant at a small distance from the ball and explosively branch into pine needle-like sparks. The fire ball is red hot and is slightly transparent. The material in the ball moves around showing that it is reacting. As the reaction of the fire ball becomes gradually weak, the sparks become like willow without branching and at last cease. The temperature of the fire ball measured by Dr. Nakaya and his colleague Sekiguchi is as follows:

| | |
|---|---|
| Before projecting sparks | 850°C |
| When the sparks are projected most actively | 930°C |
| At the end of active spark projection | 830°C |

These data derive from the average of the values of ten measurements, respectively. The sparks projected from the fire ball consist of melted and reacting matter. The reaction of the fire ball is very active in oxygen, and ceases in nitrogen. When the fire ball is blown in the stream of air, the sparks become very active at once, and from these facts we know that a proper amount of oxygen is necessary to produce the sparks.

When we analyze the matter which makes a fire ball, we cannot find potassium nitrate in it. The main components are potassium sulphide, potassium sulphate and carbon. When we heat the potassium sulphide, which is separated from the fire ball matter (ash), to a red state, it melts, becomes a red hot ball and actively bubbles. In this case the red hot ball is covered with a thin smoke layer, which is thought to be sulphur dioxide from its smell. Then the cooled and solidified fire ball matter is heated to red heat with the flame of an alcohol lamp, it becomes active again and produces sparks, but when it is inserted in the reducing part of the flame, the reaction, and accordingly the sparks, suddenly cease. It is thought that the chief components which cause the sparks may be potassium sulphide and carbon in the fire ball. If this is true, we can also artificially create the same phenomenon as Senko Hanabi, when we make a mixture of these two components, of an adequate ratio, and heat it. Accordingly, the following experiment was tried: 66% of potassium carbonate, 30% of sulphur and 4% of lamp soot (pine soot) was heated at a relatively low temperature in a porcelain crucible, being melted to make a mixture of potassium sulphide and carbon. A small amount of the product was fixed at an end of a nichrome wire and heated by the oxidizing flame of an alcohol lamp. As expected the matter became a red hot fire ball in an active state and produced sparks which were the same as Senko Hanabi.

The mechanism of producing sparks from the fire ball is not yet known, and when we display the Senko Hanabi, the fire ball is apt to fall down from the support, and the viscosity of the matter in the hot state needs also to be studied further.

Spark composition using metal: The fire branching of sparks is interesting where iron is concerned. Dr. Nakaya studied the iron sparks which are created by a grinder. In this case pure iron does not branch the sparks, but when the carbon content increases, the sparks branch twice or three times, and when the carbon content increases to more than 7%, the shape of the sparks does not change any more. Thus we can tell the amount of carbon in the iron by observing its sparks, when the carbon content is less than 7%. The iron sparks are very similar to those of Senko Hanabi.

When we use iron or magnesium powder as the ingredient for sparks, it must be coated with some resistant material to prevent it from rusting. For this purpose paraffin wax or a benzene solution of Japanese varnish etc. can be used. The examples of spark formulas are as follows:

Volcano:

|  | % |
|---|---|
| Potassium Nitrate | 55 |
| Sulphur | 9 |
| Hemp Coal | 13 |
| Coated Iron Powder | 23 |

This composition is pressed in a rather thick tube or cone with a fuse attached. It is ignited by the fuse, and produces sparks through the hole of the fuse.

Fountain:

|  | % |
|---|---|
| Ammonium Perchlorate | 70 |
| Coated Magnalium Powder | 30 |

It is used in the same way as the Volcano. The coating of magnalium must be specially perfect; otherwise the life of the composition is very short.

*Illuminant composition*

The mixture of magnesium powder and barium nitrate is usually used as it produces the most intensive light, and for fireworks an aluminum and barium nitrate composition is also used. The color of the light is generally white, but in the case of the former it appears to be rather whitish green. Magnalium is also sometimes used, but the composition sometimes burns so vibrationally, that it is not particularly recommended, though stars which burn vibrationally have been recently used to exhibit special beauty. When we use oxidizers other than barium nitrate or barium salts, the intensity of the light diminishes. Barium oxide in the flame has a strong molecular spectra and the condensed particles in the flame also emit a strong continuous spectrum. This is the reason why the light intensity of barium composition is high.

Magnesium Illuminant:

|  | % |
|---|---|
| Magnesium Powder Coated with 3 % of Paraffin | 50 |
| Barium Nitrate | 47 |
| Linseed Oil | 1.5 |
| Castor Oil | 1.5 |

This composition, which is pressed in a paper tube of about 10 cms. in diameter by a hydraulic press to a density of 2.00 grams/cm³, burns at a rate of 1.9 mm/sec, and the candle power of the light is 5000–7000 per square centimeter of the burning surface. To obtain a green light, the ingredients which produce chlorine or hydrochloric acid gas in the flame, (e.g. benzenehexachloride or vinyl chloride), is added to above formula. The amount of this ingredient should be 10–15%. This type of ingredient is also effective for the following blue and red colored flame compositions. To obtain other colored lights of high intensity, barium nitrate is replaced by potassium perchlorate, and a suitable color creating ingredient such as strontium carbonate for the red flame, sodium oxalate for the yellow flame or Paris green for the blue flame is used. In the case of blue flame the amount of magnesium in the composition should be less than 20% to obtain good color. When we use ammonium perchlorate as the oxidizer the flame color tone is excellent, but only a small amount of moisture causes reaction between magnesium and ammonium perchlorate thereby creating magnesium perchlorate which is very hygroscopic, and damages the composition. Accordingly the composition which contains magnesium and ammonium perchlorate must be kept always in a perfectly dry state. Strontium nitrate behaves as an oxidizer as well as a color creating ingredient. It is not so easy to store a composition which contains magnesium for long periods, and the pressed composition effloresces gradually on the surface which is in contact with air. To avoid this possibility the composition is pressed in a thin metal case or is packed with metal foil to protect the composition from moisture in the air. When we make stars with this kind of composition, one must not use water but rather some waterless solvent.

*Aluminum Illuminant for Cores of Stars:*

|                                | %  |
|--------------------------------|----|
| Barium Nitrate                 | 67 |
| Aluminum                       | 27 |
| Soluble Glutinous Rice Starch  | 6  |

This composition is effectively used as cores of stars. When it is used as an ordinary illuminant pressed in a container, the starch is unnecessary. The stars of this composition must be covered with another composition of high ignitability. It is difficult to obtain a good colored flame with aluminum composition.

*Sound Composition*

There are two kinds of sound composition: the report composition and the whistle composition. The composition called "Thunder" belongs to the former. The explosive sound seems to be sharpened with the rate of explosion. The relation between the nature of the explosive sound and the rate of explosion is an interesting problem, which has not yet been made clear. The description of sound as "Round" or "Sharp" may have some relation to the rate of explosion which is peculiar to its composition. The report composition must not only have a rapid rate of explosion, but also high ignitability with a fuse, which can produce the explosion easily. This type of composition generally has a high sensitivity, and accidents occur frequently with sound composition. Lately, safer compositions, which contain no chlorate as an oxidizer, have become gradually popular.

*Potassium perchlorate report composition*

It is said that perchlorates are relatively safe to handle, but ammonium perchlorate is not suitable for the report composition because of its hygroscopic nature, and only the potassium salt is used. For the heat creating ingredient only aluminum is used. If we use ferrosilicon in place of aluminum the composition does not easily explode on ignition, and even if it explodes the sound is very small. It is ne-

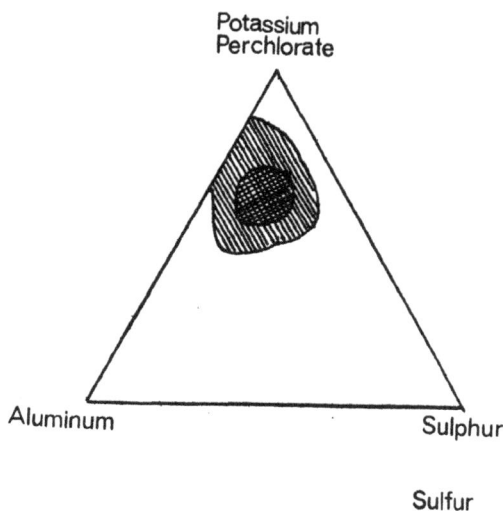

Fig. 114  Flash compositions

cessary for the aluminum to be as pure as possible and in very fine powder. As the fuel sulphur or antimony trisulphide is used, there is almost no difference between them. Only when the material has to be kept for a long time would the latter be recommended. And if potassium chlorate is the oxidizer, it would be somewhat safer to use antimony trisulphide rather than sulphur.

The intensity of the sound and the sound producing zone have been studied in a composition which consists of three components, potassium perchlorate, aluminum and sulphur, and the results are shown in the triangular diagram in Fig. 114. The shadowed part with single inclined parallel lines shows the zone in which the sound is produced, and the zone marked with crossed parallel lines shows the composition of the greatest sound. The sound zone is fairly wide. The amount of aluminum is generally increased as much as possible to obtain a strong flash. An example of the composition used is as follows:

*Thunder No. 1:*

|                        | %  |
|------------------------|----|
| Potassium Perchlorate  | 50 |
| Aluminum               | 23 |
| Antimony Trisulphide   | 27 |

When we use sulphur in place of antimony trisulphide, the sound changes very little. In effect, it is necessary to select the most adaptable and economical composition in the sound creating zone for our purpose. This kind of composition does not explode but only burns when a small amount is ignited in a free state in the air. When the amount is large, the burning changes instantaneously to detonation. The limit of the amount is likely to be about 50—100 grams. Accordingly care must be taken not to accumulate large amounts of the composition in the workroom, for the destructive power of the detonation caused by the composition is terribly large. This kind of composition is not easily exploded by friction or shock.

When this composition is exploded by ignition with a fuse, the intensity of the container or the packing has a great influence. With increasing pressure the detonation limit of the amount of powder becomes smaller than would be the case in a free state (i.e. an unpacked state), and the time interval from the ignition to the detonation will be very much shortened. This is a very important characteristic of a powder, especially in the case of aluminum composition. The composition which contains ferro-silicon in place of aluminum is not as strong as the aluminum composition. We must make the packing strong and the igniting surface as wide as possible by attaching a small amount of black powder paste to the end of the fuse in order to obtain a good report.

The influence of the loading density on the sound is not so clear. From a test, changing the density from 0.5 to 1.2, it seemed that an increase in density gave an increase in sound. A loading density higher than this value will not always give a good effect. Aluminum in fine powder gives a larger sound than when it is of large particle size, but the influence of the particle size is very small compared with that of the strength of the container in practical use.

*Potassium chlorate report composition*

The potassium perchlorate report composition which is described above is limited because a minimum quantity is required for detonation, and with less than this amounts it can not move to detonation from ignition. With very small amount it is also difficult to produce the report sound. Moreover its characteristic low sensitivity against shock and friction is sometimes not very effective for firework purposes. The potassium chlorate composition remedies the above defects, and although it is dangerous, we do still have occasion to use it.

*Red Explosive Composition:*

|  | % |
|---|---|
| Potassium chlorate | 63 |
| Realgar | 37 |

This is called "Red Explosive Composition" because of its red color, and the formula above is an example of this kind. It is very sensitive to shock and friction, and very dangerous to handle, in fact to handle it safely it is recommended that the wet process be used. The sound produced by this composition gives a finer sound wave than the potassium perchlorate composition, and is sharper than the latter. On the other hand the strength of the sound produced by this composition seems to be weaker than that of the potassium perchlorate composition. A very small amount of the red explosive composition can produce detonation. Accordingly it detonates perfectly even when the strength of the container is small, and even when the amount of the composition is very small. This composition is used for toy "Cracker Balls" etc.

*Toy Pistol Cap:*

|                                           | %  |
|-------------------------------------------|----|
| Potassium Chlorate                        | 60 |
| Red Phosphorus                            | 8  |
| Sulphur (or Antimony Trisulphide)         | 32 |

One example of this is shown above. This composition is more dangerous than the Red Explosive composition, and so sensitive to shock and friction, that it must be produced by the wet process. Red phosphorus must be free from white phosphorus and phosphoric acid. In place of sulphur, antimony trisulphide is also used without much change in the character of the composition. The composition of the above formula is used for toy pistol caps, but the one in which sulphur is replaced by antimony trisulphide is used for igniters because of its large flame.

*Thunder No. 2.*

|                        | %  |
|------------------------|----|
| Potassium Chlorate     | 43 |
| Antimony Trisulphide   | 26 |
| Aluminum               | 31 |

This formula looks like that of the potassium perchlorate composition, but this composition is more sensitive to shock and friction than the perchlorate composition. But it is used nowadays in Japan because of its low price, its low limit of the minimum amount for detonation, and its ease of detonation even in the case of a rather weak container. It is sometimes called "Flash Thunder", the name of which comes from the dazzling flash of light in contrast to the realgar composition which produces only a weak light. It is the same with the perchlorate composition, and to obtain a strong flash the amount of aluminum is increased at the expense of the oxidizer, provided that it does not disturb the production of the report. In the above formula it is also possible to use sulphur in place of antimony trisulphide.

*Potassium picrate whistle composition*
In the above instances the the sound of the explosion is used as a report. But the following composition is totally different from above, for it gives a sharp musical sound under special burning conditions.

*Whistle No. 1.*

|                     | %  |
|---------------------|----|
| Potassium Picrate   | 63 |
| Potassium Nitrate   | 37 |

The sound is like a whistle, and is called "Whistle".

The manufacture of potassium picrate: A wooden tub (30 liters) is prepared, and 20 liters of hot water is poured into it. Steam is used to bring the water to boiling point, and the steam pipe should be made of aluminum. While injecting the steam 2 kilograms of picric acid is added and completely dissolved. 600 grams of potassium carbonate is then charged, a little at a time. It reacts with effervescence, and a part of potassium picrate begins to form because of its low solubility. The end point of the reaction is determined by the sudden cessation of the bubbling. At this time the charging of potassium carbonate is stopped, and then picric acid is charged into the tub. The over-charged potassium carbonate reacts with effervescence and when the bubbling ceases, the process is finished. The liquid is somewhat acidic, the purpose of ending in the acid state being to filter the liquid easily afterwards and to avoid giving a hygroscopic character to the product. The contents of the tub are removed into another aluminum tub, which is cooled from outside with cold water, the content being stirred with a wooden stick. When it is cooled too slowly, it produces crystals which are too large, and necessitates some dangerous process such as pulverization. Thus it must be cooled as quickly as possible to obtain very fine crystals. When the cooling process is finished, the contents are removed into a porcelain funnel, which contains a filter paper. The product is spread on a paper sheet and dried in the sun. The yield is about 104 parts of potassium picrate per 100 parts of picric acid by weight. Potassium picrate consists of yellow needle shaped crystals, which are somewhat difficult to dissolve in water. Its solubility in water is about 6 grams per liter at normal temperature. Potassium picrate is rather sensitive to shock and friction, but the degree is not so high as the Red Explosive Composition. Even when it contains a small amount of moisture, it explodes. The burning rate in the air in the free state is terribly fast, and great care must be taken when handling it. The picrate should only be prepared in small quantities when it is necessary. The manufacturing process is very simple, but some skill is required. Care must be taken that the abandoned mother liquor of the reaction does not pollute the effluent system.

The standard formula is given above as Whistle No. 1. Less potassium nitrate causes an explosion and more produces no whistle. Potassium picrate and potassium nitrate are mixed by a hair sieve in accordance with the mixing process for dangerous compositions described above. This kind of composition seems safer than potassium picrate alone, according to the results of the drop hammer test.

*Gallic acid whistle composition and other whistle composition*
The standard formula of the composition which contains gallic acid
is as follows:

*Whistle No. 2.*

|                       | %  |
|-----------------------|----|
| Gallic Acid           | 25 |
| Potassium Chlorate    | 75 |

This composition is very sensitive to shock and friction. The degree
of danger in handling it is almost the same as in that of Red Explosive
Composition. The mixing process is the same as that of Whistle No. 1.
When more than 15 % of water is added to the composition, it becomes
less sensitive.

As a safer composition a mixture of salicylate of sodium and potassium perchlorate is recommended. The whistle is somewhat smaller
than that of the above composition, and this composition must be
protected from moisture, because it is very hygroscopic.

Whistle composition is generally pressed into a paper tube leaving
a half of its length unloaded. It is very important to press the composition firmly to avoid an explosion when it is burning. The paper tube
should therefore be inserted in a mold and the composition pressed
in it several times with a hand press. This process should take place
behind a defense plate. The composition is carried in a safe container
in small amounts from another room to the workroom to avoid a large
unexpected explosion. An example of the product is shown in Fig. 115
The paper tube should be made of dense material. The ratio of the
diameter of the tube to the length should be less than 1:3, but when
the tube is too long, it cannot produce a good effect, i.e. the larger the

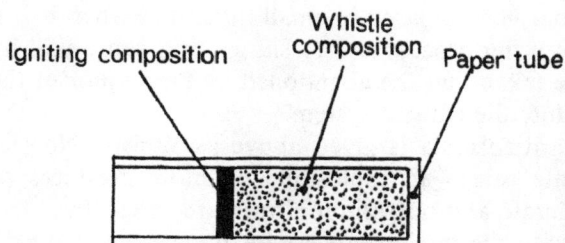

**Fig. 115**   Firework whistle

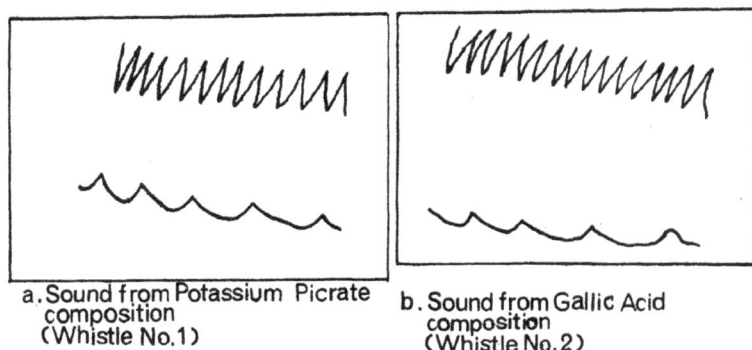

a. Sound from Potassium Picrate composition (Whistle No.1)

b. Sound from Gallic Acid composition (Whistle No.2)

**Fig. 116** Sounds

the diameter, the louder the whistle and the tone decreases accordingly. In the case of a very small tube of 3 mm. in diameter the potassium picrate composition will not whistle, and only the gallic acid composition is usable. Fig. 116 shows the wave forms of whistle from Whistle No.1 and Whistle No.2. The frequency of both sounds is about 2600 per second.

*Smoke Composition*

Pyrotechnic smoke for practical use can easily be obtained from a simple smoke device without any special technique, and must be safe to handle. Here, only the smoke which is used widely at present for fireworks is discussed.

The processes are classified as follows:

1. The burning product grows to form a mist which absorbs the moisture from the atmosphere. For example the hexachloroethane smoke composition creates a vapor of zinc chloride or aluminum chloride, which becomes white smoke when it combines with moisture in the air. At present carbon tetrachloride smoke composition is seldom used.

2. The material is first vaporized by the heat of combustion and then condenses again to fine solid particles which create smoke. The white smoke caused by sulphur and yellow smoke caused by realgar belong to this type.

3. Smoke caused by imperfect combustion of the composition. This type of smoke is widely used as black smoke in fireworks. It utilizes the carbon particles created by the imperfect combustion of naphthalene or anthracene.

4.   Colored smoke produced by vaporizing dye. A volatile dye is first vaporized and then it condenses to fine solid particles, which look like colored smoke. Generally this kind of smoke shows its own beautiful color by the reflected light, but it looks dirty when the light is weak. In a cloudy sky it does not have a good color. Generally the color of the water solution of dye from permeated light is the same as the color of cloudy solid particles in reflected light. Physically it is an interesting problem.

*Hexachloroethane smoke composition.*   This composition consists of three components, hexachloroethane, zinc and zinc oxide. The burning product of this composition is mainly the vapor of zinc chloride, and the temperature of the burning reaction may be about 900 °C. The composition is generally contained in a perfectly moisture-proof tin. The burning time of this composition is shown in Fig. 117 as an example. The dimensions of the smoke container were 80 mm. in diameter and 114 mm. in height.

The influence of the quality of zinc oxide upon the burning rate of the composition is remarkable. Three kinds of zinc oxide were used for a composition of 50% hexachlorethane, 28% zinc dust and 22% zinc oxide, and tested to give results, as follows:

| Class of Zinc Oxide | A | B | C |
|---|---|---|---|
| Duration (sec) | 227 | 267 | 347 |
| Color | greyish | greyish | pure white |

The quality of zinc oxide of class C is better than that of class A, and

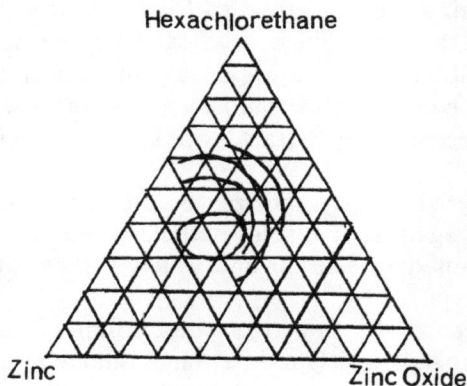

Fig. 117   Smoke compositions

the quality of class B is better than that of class C. In short, the better the quality, the longer the duration. Zinc of good quality makes the color of the smoke pure white.

The quality of zinc dust and the particle size of hexachlorethane did not show such a marked influence upon the burning time. A particle size of hexachloroethane of less than 2 mm. can be used successfully.

The mixture of zinc dust and hexachloroethane reacts violently, when water is added to it drop by drop. It should be noted that this kind of smoke sometimes causes fire in the presence of moisture. This composition never explodes with shock or friction like other explosives. The so-called "Explosion of Smoke" seems to be the explosion of an aluminum composition, which is contained in the igniter for the smoke, or the explosive bursting of the tin by interior expansion, caused by the heat evolved when the water entered the tin.

The smoke composition is so perfectly sealed in a tin, that it must be designed specially to ignite without fail. Generally a brass cap, the thickness of which is less than 0.2 mm. is soldered into the tin and a quantity of thermit is charged into it, on the top of which an igniting star is inserted as shown in Fig. 118. When the head of the igniting star is rubbed with a wooden striker, on which a composition of red phosphorus and antimony trisulphide is pasted, ignition takes place. Next the thermit is ignited, generating a large amount of heat which melts the brass cap and the hot melted metal drops on the smoke composition directly and infallibly igniting it.

The manufacturing process for smoke composition is not dangerous, but care must be taken to reduce the moisture in the ingredients as far as possible and frequently the temperature of the composition during mixing should be tested by hand or by some other method. During the

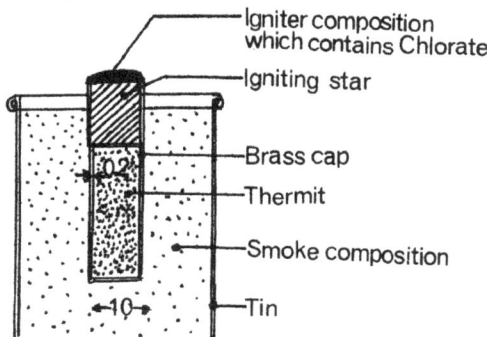

**Fig. 118** Smoke cannister

mixing process the light zinc oxide is first charged into the mixer and the heavy zinc dust is charged last. When a rotary mixer is used, the number of revolutions per minute shall be less than 20. It takes about 20 minutes to mix 100 kilograms of the composition. The volume of the mixer is designed so as to have 2/3 of the total volume empty with the volume of the charge occupying 1/3, the apparent specific gravity of the smoke composition being 1.1—1.5.

The composition is sealed in the tin with a canning machine, but it is rather different from the canning of food and more difficult especially with a large smoke. The mouth of a can must be an exact circle and must not be warped by the contents. The grove of the cap is filled with gum solution before the canning to keep it airtight.

There is another kind of composition in which the zinc powder is replaced by aluminum powder. This composition creates greyish smoke and is not often used in Japan.

*Sulphur smoke composition*

The smoke is white and easy to generate, and this composition does not react with water and is not hygroscopic. Therefore it is used widely in fireworks. The smoke consists of solid particles, and is not affected by the moisture in the air.

An amount of potassium nitrate and the same amount of sulphur are mixed well together. Moreover a small amount of realgar is added to it to ignite easily and to smoke smoothly.

*Sulphur White Smoke:*

|                    | %    |
| ------------------ | ---- |
| Potassium Nitrate  | 48.5 |
| Sulphur            | 48.5 |
| Realgar            | 3.0  |

This composition is pressed firmly into a paper tube, both ends of which are sealed with gypsum, and a hole is bored into the composition on the side near to one end. A piece of quick match is inserted into the hole. When it is ignited a white smoke generates through the hole. The hole must not be too large or else the smoke will burst into flame. This composition can also be used for white smoke stars, when it is mixed with soluble glutinous starch and shaped into cylindrical stars, each of which is covered with gypsum, and then a hole is bored into this gypsum for the smoke to come out. Care must be taken not to cause ignition during the drilling operation. A small amount of hemp coal is sometimes added to the composition for good ignition.

*Realgar smoke Composition*

A composition which consists of potassium nitrate, sulphur and realgar can produce smoke of white, light yellow, deep yellow or purplish grey according to its formula. The yellow smoke is the most popular, and an example is as follows:

*Realgar Yellow Smoke*

|  | % |
|---|---|
| Potassium Nitrate | 25 |
| Sulphur | 16 |
| Realgar | 59 |

In this formula the percentage of sulphur is rather small. The method of generating smoke is the same as that described in (2). The moisture in the composition seems to have a little influence upon the burning rate of this composition. In the case of smoke stars for shells, which are used for willow or chrysanthemum, the sulphur is increased much more to make the burning time shorter. For this purpose an amount of sand is sometimes added to the composition, or the stars are covered with gypsum so that they burn quickly under pressure as described in (2). An example of the formula for stars used in a willow shell is as follows:

*Yellow Willow:*

|  | % |
|---|---|
| Potassium Nitrate | 43 |
| Sulphur | 10 |
| Realgar | 37 |
| Hemp Coal | 4 |
| Soluble Glutinous Rice Starch | 6 |

The following composition is used for daylight shells. The smoke looks white. This may be only a smoke from a burning process, and it may differ from the smoke generators mentioned above.

*White Chrysanthemum:*

|  | % |
|---|---|
| Potassium Nitrate | 66 |
| Realgar | 13 |
| Lamp black | 5 |
| Hemp Coal | 5 |
| Soluble Glutinous Rice Starch | 11 |

The volume of lamp black is so large that a little amount of starch is used as the binder. The distinguishing feature of this formula is that it has no sulphur.

### Anthracene or naphthalene smoke composition

When a carbon rich compound like anthracene or naphthalene mixed with some oxidizer is burnt, it creates carbon particles due to incomplete burning. (If the amount of oxidizer is too large, a good smoke is not produced).
For example:

### Anthracene Black Smoke I:

|                        | %  |
|------------------------|----|
| Potassium Perchlorate  | 56 |
| Sulphur                | 11 |
| Anthracene             | 33 |

This composition is filled, pressed into a tin and ignited through a small hole, when a flame will blow out of it. A thick black smoke is generated from the flame. This composition is different from the others, because the flame appears first and then generates smoke from it. (In many other types the production of flame generally disturbs the smoke). The temperature of the flame may be about 800 °C. This smoke is adaptable for a black "Dragon" with a parachute, but this composition is a little difficult to ignite and the burning is only stabilized under a pressure which is maintained by the smoke case. It is not easy to ignite this composition, even when it is in a powdered state, and a special method must be applied e.g. using thermit. The composition is safer to handle than the following composition, Naphthalene Black Smoke I, and care must be taken because of its explosive nature.

### Naphthalene Black Smoke I:

|                              | %  |
|------------------------------|----|
| Potassium Chlorate           | 44 |
| Antimony Trisulphide         | 24 |
| Naphthalene                  | 26 |
| Soluble Glutinous Rice Starch | 6 |

This composition is used to make stars for producing a beautiful black smoke, but it is also more explosive than Anthracene Black Smoke I. Naphthalene sublimes out of the star when it stands for a long time leaving behind potassium chlorate and antimony trisulphide which increases its explosive nature. In spite of this disadvantage it is still used widely in Japan, because it is the most beautiful of all black smokes, but it is better not to use this composition, if possible.

*Naphthalene (Anthracene) Black Smoke II:*

|                              | %  |
|------------------------------|----|
| Hexachloroethane             | 62 |
| Magnesium                    | 15 |
| Naphthalene (or Anthracene)  | 23 |

This composition has no oxidizer. The Hexachloroethane and magnesium react with each other producing a smoke of magnesium chloride, and the heat developed causes the isolation of carbon particles from the naphthalene, thus making the color of the smoke black. This composition is loaded into a tin and used as a signal.

*Anthracene Black Smoke II.*

|                               | %                       |
|-------------------------------|-------------------------|
| Potassium Perchlorate         | 57                      |
| Anthracene                    | 40                      |
| Hemp Coal                     | 3                       |
| Soluble Glutinous Rice Starch | 7 (additional percent)  |

This composition is a modification of the composition Anthracene Black Smoke I for stars, and it is more ignitable than Anthracene Black Smoke I.

*Colored smoke composition with dyestuffs*

Dye smokes show their natural color by reflected light. The adjustment of color is achieved by the same principle as that of mixing dyestuffs for printing. A colored smoke composition consists of dyes, a small amount of oxidizer and some kind of carbohydrate, which adjusts the burning temperature. The composition is pressed into a case or is made into smoke stars, adding a binder (e.g. glutinous rice starch). Potassium chlorate is used as the oxidizer. To adjust the temperature and provide a fuel, starch or wheat flour is used. Sugar, or milk sugar, creates very beautiful colored smoke, but it is not widely used in Japan because of the rather high cost.

Dextrin is also a good ingredient for this purpose. The creation of flame disturbs the generation of smoke, and so the smoke must issue out of a small hole in the container after being rapidly cooled, or, in the case of stars, they must be projected rapidly through the air to extinguish the flame. The burning temperature of this type of smoke composition is about 400 to 600 °C, and some part of the dye may be damaged at this temperature. If the temperature rises to a higher level than this during the burning process, the smoke becomes colorless. A few examples of formulas for colored smoke compositions are shown as follows:

*Red Smoke:*

|                      | %  |
|----------------------|----|
| Potassium Chlorate   | 25 |
| Rhodamine B          | 24 |
| Para Red             | 36 |
| Wheat Flour          | 15 |

*Blue Smoke:*

|                      |    |
|----------------------|----|
| Potassium Chlorate   | 28 |
| Methylene Blue       | 17 |
| Indigo Pure          | 40 |
| Wheat Flour          | 15 |

*Green Smoke:*

|                      |    |
|----------------------|----|
| Potassium Chlorate   | 28 |
| Auramine             | 10 |
| Methylene Blue       | 17 |
| Indigo Pure          | 30 |
| Wheat Flour          | 15 |

*Violet Smoke:*

|                      |    |
|----------------------|----|
| Potassium Chlorate   | 26 |
| Indigo Pure          | 22 |
| Rhodamine B          | 16 |
| Para Red             | 21 |
| Wheat Flour          | 15 |

In the case of Red Smoke if the amount of Rhodamine B is increased, the smoke becomes reddish violet and, when decreased, it becomes orange red. In the case of Green Smoke, Auramine is somewhat hygroscopic, and it is better to protect the composition from moisture.

To increase the burning rate of smokes some people often add sulphur to the composition, but it is not recommended in view of the increased sensitivity. To increase the burning rate, it is better to widen the burning surface by granulating the composition with the smallest dimension of the grains determining the burning time.

Potassium nitrate and potassium perchlorate are not such good oxidizers as potassium chlorate, because they cannot create colored smoke so smoothly as potassium chlorate and the percentage in the composition must be greater than that of potassium chlorate. Other heat producing materials such as celluloid and nitrocellulose are also used in place of oxidizer and fuel, but the ratio of the amount of this material to the amount of dye must be increased more than the ratio in the above formulas, and the volume of the smoke decreases accordingly. Guanidine nitrate is also used as a low temperature burning material and is combined with other heat producing materials e.g. celluloid, nitrocellulose, to protect the dye from attack of heat and to keep the ash porous. For example a mixture of 40% guanidine nitrate, 35% celluloid powder and 25% dye makes a good colored smoke except in the case of a dye with a high sublimation point like indigo. This principle is also applied to insecticidal smoke. Nevertheless the method of using heat producing materials in place of oxidizer and fuel is not so adaptable for fireworks because of the rather thin smoke, but it is recommended for some large signals of long duration.

*Powder Pasted Paper*

The powder pasted paper is made of Japanese paper with some kind of composition pasted on it, and is used widely as fuse and igniting material. The paper, on which colored flame composition is pasted, is used especially for the shell "Falling Leaves".

The kind of the composition for powder pasted paper depends on its use, and it must be selected according to the purpose.

*Black powder pasted paper:* 4 parts of black powder is mixed with 3 parts of water in weight ratio. In winter potassium nitrate easily oozes out of the pasted black powder, and we must reduce this escape by heating it with some safe form of heat. Sometimes soluble glutinous rice starch is added to the paste to give it adhesive power, but when we use grains of gunpowder which are manufactured by milling, the adhesive power is so great, that there is no need to add starch to the paste. Newspapers are fixed on a work table as a protective cover and Japanese "Kozo" papers are spread out on them. Then the black powder paste is pasted on them uniformly with a brush of good quality. The pasted paper should be dried on a drying frame in the sun, and for manufacture a bright, clear and calm day must be chosen. Generally infra-red driers or other heat sources do not give as good a result as the sun because of the lack of uniformity in drying. When pasting, the powder paste must be stirred well, continuously, to prevent the components separating from each other. The quantity of the composition pasted on one side of the paper is 0.01—0.02 gram/cm² for fuse,

0.01 gram/cm² for ordinary use and 0.02 gram/cm² for special ignition material. The thickness of the powder pasted with the quantity 0.02 gram/cm² is rather large, and to obtain such a thickness it must be pasted on one side two or three times. For use as a fuse, paper pasted on both sides is generally used, but for other purposes paper pasted on one side is used. Test pieces (10 mm. in width) taken from a thick paper, pasted on both sides with gunpowder, showed the burning rate of 1.7 cm/sec. in the atmosphere.

*Red Thermit Paper:* The composition of minium and ferro-silicon, (e.g. in a ratio 4:1) called "Red Thermit", is mixed to a paste with 5% celluloid solution in amyl acetate. It is pasted on a Japanese "Kozo" paper like the black powder pasted paper. It is preferable to dry it in the sun for to use a drier is sometimes dangerous, because the vaporized gas of amyl acetate may come into contact with an overheated part of the drier and cause an explosion. Thermit paper is easily ignited by friction, and care must be taken in handling it. This paper can be effectively used for igniting smoke compositions which are difficult to ignite. In this case the thermit paper is used together with a black powder paper, one on top of the other, or sometimes by pasting red thermit on a black powder pasted paper. To adjust the burning rate of the paper the ratio of the amount of minium to ferrosilicon can be changed, When the value of the ratio is decreased the burning rate is delayed.

*Colored fire paper:* This is manufactured the same way as black powder pasted paper using a colored flame composition which is used for stars. Sometimes, immediately after pasting the composition on the paper, the pasted composition is covered with another paper to hold the composition well between the two papers. When it is necessary, it is manufactured in several layers like a sandwich, and is used as a part of the shell "Falling Leaves".

*Potassium Nitrate paper:* A solution of potassium nitrate in water is spread on a Japanese paper and dried. This paper is used for igniting or for helping the ignition of some fireworks. If letters or pictures are drawn with the solution on paper, they disappear when it is dried, but when it is ignited at one point of the painted part, the fire runs only on the drawn line, and the original letters or pictures appear again. This is used as a toy. Hidden pictures can be made also with other oxidizers.

Powder pasted papers often produce fire when they are cut and are very dangerous, and the cutting of a number of multifolded powder pasted papers at one time should be forbidden. Be sure to cut them in small quantities with a sharp edged tool, and don't leave the cut pieces near the cutter during the operation, i.e. the pieces should be quickly removed from the cutter and the cutting place, and they must be kept in an incombustible container which must be covered with a lid. The process is successfully operated by two workers working together. There are examples of accidents caused by cutting black powder pasted paper, and the author has had an experience of fire when cutting red thermit papers.

*Fuses* The purpose of the delay fuse for fireworks is to transfer fire and to ensure a time delay. It differs somewhat from industrial fuse.

*Handmade fuse:* A powder pasted paper, which is pasted with black powder on both sides, is cut to a size (250 mm. × 75 mm). A quantity of gunpowder powder is prepared, crushing the grains which are moistened first. The paper is then sprinkled with the powder uniformly. A long and slender bamboo stick (3 mm. in diameter) is placed on one edge of the longer side of the paper and the paper is rolled on it by hand. This rolled powder pasted paper becomes the core of the fuse. A Japanese "Kozo" paper (360 mm. × 270 mm) is attached with its one shorter side edge on the rolled core paper and rolled on it, so that the paper winds the core as a cover. It is rolled on a wooden work plate with both hands until it is firmly wound, and then the bamboo stick is pulled out and removed. The first securing operation is applied to this prewound paper as follows: after being put on a thick hard wooden workplate, it is pressed and rolled with another wooden plate (450 mm. × 90 mm. × 32mm), which is held by both hands repeatedly. In this case caution must be taken for heat is sometimes generated. At last the paper is firmly wound like a stick, or a pencil with a thick lead. With the hand pressure of an average man it is necessary to repeat the rolling about 100 times with the rolling plate. Another Japanese "Kozo" paper of the same size is wound again on the rolled stick as before, so as to reinforce the cover of the core. The starting end of the paper is fixed on the stick with paste. The same rolling process is done as a second securing operation. Then the third securing operation is applied the same way as the second. At last the end of the paper is fixed on the stick with paste and the product is dried in the sun. Fuse made by means of this process is 6.5 mm. in diameter and its core of black powder is 3.5 mm. in diameter. The burning rate of this fuse is about 1.08—1.25 mm/sec, and the deviation is rather wide. When the firework shell explodes in the neighborhood of the highest point of the trajectory, a deviation of such a degree may be allowed because the shell travels slowly when higher. When a hand rolled fuse is compared with an industrial fuse, the ignitability and ignition power of the former are better than that of the latter: the former can produce a longer flame than the latter, and on chopping or peeling, the former can hold its powder core intact, and this character enables us to enlarge the ignition power and ignitability, which we cannot do with the industrial fuse. Nevertheless the former is hand-made, expensive and has rather large time deviations. It is therefore preferable to use a machine-made fuse, but it must be specially made, e.g. the cover of the core powder must be free from tarry matter, because the core powder sometimes absorbs the tar from the tarred covering material and it makes the fuse unignitable.

The fuse made by means of the process described above is a normal one, and for 10 and 12 inch shells or furthermore larger shells the diameter of the core and the thickness of the cover are enlarged even more.

The process described above is for the main fuses of firework shells, but for other uses it may be simplified. As a powder core only the powder of gunpowder is used instead of powder pasted paper. The manufacturing process in this case may be very simple.

The hand rolled fuse made in the way described above, has rather wide variations and care must be taken in the way fuse is selected for timed fireworks. For example the shell "Five Reports" requires five accurate time intervals and so all five pieces of fuse must be cut from the same length, and care must be taken to separate different batches of fuse.

"Dark Fuse" which creates no visible sparks in the distance during burning is also made by means of the same process except that other core compositions are used, e.g.:

| Dark Fuse: | No.1 | No.2 |
|---|---|---|
| | % | % |
| Potassium Nitrate | 36 | 56 |
| Realgar | 45 | 34 |
| Paulownia Charcoal | 10 | 10 |
| Sulphur | 9 | — |

The fuse which contains realgar must not make direct contact with chlorate composition.

*Fuse for toy fireworks:* The fuse must be easy to ignite with a match; it must be easily ignited by a flame and not have a thick cover.

In Japan a twist of paper which contains black powder is widely used because of its cheapness. Potassium nitrate paper is effectively used in this case.

*Stars*

The stars may be divided into two broad classes: one group consists of light stars and the other of smoke stars. The stars for night use belong to the former and the stars for daylight use belong to the latter. But there are exceptions: a special intensive-light star sometimes used in daylight, and a smoke star which is sometimes used at night, being illuminated by other light stars to give a special mysterious effect.

Furthermore stars are divided into other classes by form (namely cubic stars, round stars and cylindrical stars), the manufacturing processes of which are different from each other.

*Binder for star compositions.* In general stars are formed out of some colored fire or smoke composition, in which a quantity of binder is mixed. In this case water or other suitable solvent is added to it. It is also possible to form the composition in a dry state by pressing, without the binder, but pressing influences the property of the stars, and in practice the pressing process is laborious and troublesome. The weak point of the star making process by using a binder is that after the stars are formed they must be dried, warmed or given an aging time. However this method can be applied to various kinds of compositions, requires no special forming machine and is not so laborious. There are two kinds of binders, water soluble and those insoluble in water. In Japan the former is generally used and the latter is used only in special stars in which water is impossible. In order to dry stars in a short time a water insoluble binder with volatile solvent is sometimes used.

Soluble glutinous rice starch is mostly used as a water soluble binder in Japan, and it can be mixed in a composition as a powder. The composition which contains the starch becomes a paste when we add only an amount of water to the composition and knead it by hand. The process of the pasting is so simple and the starch contains no acid matter, so can be safely used. The amount of the starch in a composition should be small, just enough to solidify the stars. When there is too much starch, it makes stars unignitable or lowers their burning temperature, Generally the starch unfavorably influences the burning characteristics of the stars. The amount of starch required for solidifying the stars is not fixed definitely, but varies according to the kind of composition. For the black powder type spark composition or colored flame composition the amount is 2—6%, and for aluminum composition it is necessary to increase the amount to 7—10%. Round stars containing aluminum are especially difficult to form when the amount of starch is too small.

Gum arabic and dextrine are also soluble in water, but the viscosity of the solutions is so small, that it cannot be compared with that of glutinous rice starch. For example when 20% solutions (20 grams of the matter in 100 cc. of water) are compared, gum arabic or dextrine makes a solution of comparatively low viscosity, but the glutinous rice starch is gelled without any fluidity. The nature of the glutinous rice starch enables us to solidify star compositions with the smallest possible amount of extra matter, and in Japan the starch is mostly used for forming stars under the name "Mizinko", with little use being made of gum arabic or dextrine.

By an old process, in place of glutinous rice starch ordinary rice starch is used as follows: The rice is first boiled to a paste, and then mixed well with the composition. Then the pasted composition is dried and ground to powder in a mill. This powder can be made into stars by kneading it with a quantity of water. It is said that ordinary rice starch gives a smaller amount of cinder than glutinous rice starch, but the author has not been able to find such a difference. The residual ash after a star is burnt, can be produced not only by the binder, but also by the other components.

As a water insoluble binder a solution of shellac in alcohol is recommended. Shellac as a binder is used for relatively large stars, e.g. of 20—30 mm. or more in diameter, and it would take about four days to dry stars of about 40 mm. in diameter. When the stars contain hygroscopic materials, absolute alcohol is used. The amount of shellac which is necessary as a binder is roughly 5% or more of the composition.

Other kinds of binders such as a solution of celluloid in amyl acetate or collodion can be used, and the drying time is very short in this case. Linseed oil may also be used, and in this case the composition may be solidified without drying, but care must be taken that the oil does not damage the ignitability of stars. This method is not used in Japan except in special cases. Plastic binders are not used widely in Japan because of the manufacturing problems. The poisonous vapor, the heat of polymerization which sometimes causes fire, and the adhesion of the material to the container prevent its application to our firework processes.

**Fig. 119**  Star making tools

*Light stars*  The manufacturing process, using soluble glutinous rice starch as the binder, takes place in the following way:

*Cubic stars:*   An amount of composition with the binder "Mizinko" is placed in a bowl and a definite amount of water is poured into it. When the quantity of water is too great, it is difficult to cut the kneaded mass because of the adhesion of the composition to the tools, and when the amount is too small, the composition cannot be solidified because of an insufficiency of adhesive power. It is better in fact to work out before manufacture the amount of water required for each composition or to use the following method: A small part of the composition is taken out, and an amount of water is added to the remainder of the composition which is then kneaded into a paste. The former part of the composition is then added in small quantities, kneading and adjusting the viscosity at the same time. When the kneading is insufficient, the composition is not uniform and creates a lot of cinder when it burns. The kneaded composition in the bowl is covered with a lid so that it does not dry out. The tools for cutting the kneaded composition are prepared as in (Fig. 119).

First a thick wooden chopping-board (30 cm. × 60 cm. × 3 cm.) is prepared and around its edges are four thin, long, wooden edge-plates, the thickness of which is the same as that of the cubic stars, the edge-plates are fixed with small nails or brass screws. The kneaded composition is then placed on the chopping-board between the edge-plates, pressing with the fingers, so that the surface of the pressed composition is little higher than the edge-plate. Using a wooden hammer the surface of the composition is patted until it becomes as high as the edge-plates, and then a thin and long edged knife is pushed on and along the edge-plates from one side to the other of the chopping-board, removing the excess composition. The four edge-plates are then removed, so that a sheet of composition of uniform thickness remains on the board. The sheet is cut checkerwise into squares as follows: First the sheet of the composition is sprinkled with powder of the same composition and with a knife it is cut into strips which are the same thickness as the sheet of composition, thus producing a row of several, exactly square, sticks of the composition. Each stick is rolled 90°, so that the new surface produced by cutting is uppermost, and the old surface on which the composition powder was sprinkled is at the side, to prevent the sticks from adhering to each other. The cut pieces are removed to the other side, being slid along the board as they are cut, to produce a row of paralled strips. The strips are brought together and turned 90° so that they are in close formation. These pieces are cut with the knife at right angles to the strips, the same width as the thickness of each strip. Thus we have regular cubes of composition. The cubes are removed from the chopping-board into a sieve so that small particles of composition can be sifted out and the cubes which stick to each other can be separated. The size of the cube varies according to use. For example a cube with a 3 mm. or 6 mm. side is used for the core of round stars, while a 12 mm. cube is used for smoke stars. The stars are dried in the sun, but the stars which are relatively large are best dried in the shade, because stars which are dried in the sun immediately after the cutting are likely to dry only on the surface, with the result, that the inner part of the cube is not easily dried, and finally it cracks on the surface. A worker should be able to make 30,000 3 mm. or 15,000 6 mm. stars in a working day. When the cubes are covered with an igniting composition in the way described later, they can be used as complete stars.

*Round stars* (*color changing stars*)   Round stars are made by adding composition layer by layer, so that they are enlarged into balls of the proper size, and the cores are generally small cubic stars. The core has a 3mm. side for a completed ball less than 15mm. in diameter, 4.5mm. for a ball of 15—18 mm. and 6mm. for a ball of more than 18 mm. In place of cubic stars some use rape-seed or cereal-seed, and sometimes shot for the cores. The rape-seeds are so light and so small, that very much time and labor is needed to enlarge them without them sticking to each other at the beginning of the manufacturing process, so they are not much used at present, but it is said that using such unignitable matter for the cores makes the end of the star-light quite clear, and the artistic view is excellent in the chrysanthemum shells.

The manufacturing process of round stars from cubic cores is summarized as follows: The bowl for forming the stars has a spherical bottom and is made of aluminum or other light metal; in fact any bowl of 30 cm. in diameter sold on the market for cooking is suitable for manufacturing stars by hand. For manufacturing a larger quantity of stars at one time it is better to use a larger bowl which has a base which is wider and flatter. In this case the bowl is hung from the ceiling with ropes and operated by hand. Compositions are prepared according to the color changing plan. A mushy paste of a composition, which is the same as the composition for sprinkling is prepared. One part of the composition is mixed with water in another small bowl and stirred to a paste. Fireworkers in Japan call this mushy paste "Toro", and the viscosity of the paste is very important. When it is too high, the cores stick to each other in groups of two or three, especially in the case of light cores, and when it is too low, the cores grow so slowly that the water in the Toro soaks into the cores. The viscosity of the Toro varies very much according to the type of composition, so the amount of water in the Toro must be adaptable for each composition, especially in the case of a composition which contains aluminum powder which has a very low viscosity. Then it is necessary to add more glutinous rice starch than one would in ordinary compositions. In a word, the viscosity of the Toro should be as high as possible provided that the manufacturing process is possible. The development of the core into a star is shown in Fig. 120.

A correct quantity of the cores are put into the pasting bowl and a small amount of the Toro paste is poured over them; then the bowl is shaken round by hand or by other means until the paste completely and uniformly covers the core grains. A small amount of powder, which consists of the same composition as the paste, is sprinkled on

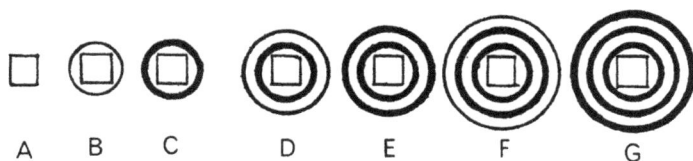

**Fig. 120** Stages in the manufacture of round stars

the pasted grains and the bowl is shaken round again. When the amount of the powder is correct, it is all fixed on the grains and the bottom of the bowl is so clean that it glistens. When the quantity of paste is too large, the paste adheres to the inside of the bowl and prevents manufacture, and when it is too small the powder remains in the bowl without fixing to the grains. The process of covering the grains with the paste, sprinkling the powder on the grains and shaking, is repeated two or three times, and the grains gradually change shape from cubic to round. Once they have been dried completely in the sun, and the operation is repeated, the grains become an almost perfectly round shape, namely A becomes B in Fig. 120. When the composition of the core is "Flash", it is difficult to ignite, and in this case the composition of the powder and the paste must be different from that of the cores for ease of ignition. This kind of composition is called "Cooperative powder of Cores". Next a composition of very weak light intensity is layered on the grains. This layered composition is called "Changing Relay". This process is done once or twice, so that grains become like C. The purpose of the Changing Relay is to make the stars seem as if they change color simultaneously and clearly in spite of their irregularity, though it only appears to be so, because the light intensity of the stars is weakened between the two layers. The grains are then coated with a layer of the next composition by repeating the same process. After coating two or three times the grains grow like D, and they are well dried in the sun. The grains are coated with the Changing Relay again like E and the next composition is layered on the grain like F. A thrice color changing star is shown in Fig. 120.

In the case of color changing stars, what kind of composition should be layered inside and what kind of composition should be layered outside a star construction? As a general rule a dangerous composition is layered inside and safer composition is layered outside, or, an expensive composition is layered inside and a cheap one is layered outside. In the cores it is preferable to use a composition which produces light which ends clearly in a flash, so, taking into consideration the nature of the trajectory of the stars, a composition of high specific gravity is layered inside and one of low specific gravity is layered outside.

Lastly black powder type compositions are layered on the grains as at G, and the stars are complete. Chrysanthemum 8, the burning rate of which is relatively slow, is first used, and Chrysanthemum 6, which has a faster burning rate than that of Chrysanthemum 8, is used next.

Thus the slower burning composition is layered first and the faster one next, while the fastest, (i.e. blackpowder) is sprinkled on the grains in the finishing process. The method outlined above is the usual process, but in the case of stars which are difficult to ignite such as those of aluminum a specially prepared composition is first layered on the grain before the black powder type compositions mentioned above, in the finishing process. The problem of ignition is very important, not only with round stars, but also with everything else, and a new device for igniting may create new kinds of stars.

The speed of manufacture of stars is closely related to the drying time and the state of the grains to be dried. When the dried grains are again pasted and sprinkled with the next composition, the moisture soaks into the inner layer of the grains which have already dried. It soaks gradually deeper and deeper towards the core of the grain as time passes, and it is necessary to operate quickly, and at once put the grains to dry. The rate of soaking of the moisture varies according to the kind of composition, and it depends especially upon the amount of glutinous rice starch in the composition and the temperature of the stars. When the amount of the starch is too great, there is not very much effect in a short time, but if the grains are left without drying for a long time, the moisture soaks into the center of the grains. Thus the moistened grains make a hard dried layer on each surface during drying, and the center of each grain cannot be dried any more. We call it "Driven In". The composition which contains a relatively small amount of glutinous rice starch does not cause such a phenomenon, for in this case though the moisture soaks into the center of the grains easily during every pasting process, they can easily be dried out during every drying process (i.e. the moisture is driven out easily through the layers of compositions). The limit of the amount of starch (e.g. binder), which does not cause the "Drive In" phenomenon has not yet been systematically studied, but in the case of ordinary potassium perchlorate colored flame composition or spark composition containing aluminum it may certainly be about 6%. In the case of aluminum composition the viscosity of the paste is too low to produce the grains. When the temperature of the grains is low, the rate at which the moisture soaks into the grains can be made slower, so it is better to leave the grains in a cool place after drying by heat and cool the grains sufficiently before the next coating operation.

The thickness of the layer that is added between the two drying steps depends upon the weather. In summer time on a clear hot day it should be rather thick, but in winter time on a cloudy day it should be rather thin, though in the case of a composition which contains a relatively large amount of glutinous rice starch and which has a high viscosity, success cannot be achieved by this procedure, and the relation between the thickness of the layer and the weather will be opposite to the above. In this instance, when the grains are dried in the strong sunshine and the new layer, coated on the grains, is too thick, only the surface of the layer is dried, and the inside cannot be dried easily. On the other hand even when they are dried on a cloudy day with a rather thick layer, the inner and outside part of the grains are dried successfully and uniformly, because the moisture soaks into the grains very slowly in this composition. Thus the thickness of each layer on the grains depends upon the weather and the nature of the composition, and is usually about 0.5—1.0mm. on average, though in black powder type composition the thickness is about a half of this. On a day with a clear blue sky the coating process is done three times a day in May, four or five times a day in March, and seven or eight times a day in June and July in Japan. The coating process must be finished by two or three o'clock in the afternoon in order to dry the grains during the day time.

The above process depends very much upon the weather and it is rather difficult to work out a strict program. Accordingly it is useful to use a drier, (e.g. of infra red rays) to dry the grains as an auxiliary operation. A drier, where the heater is placed under the drying frames, cannot be used to dry star grains because of the danger involved.

The complete round star grains, even when they are made very carefully, tend to have some small deviation in their dimensions. For example, a sample of the variations in the diameters of the grains is shown as follows. The values show the measured diameters of the grains. The letters S, K, A, B and C show the kinds of stars and the additional figures to the letters are the named diameters of stars.

| S15 | K12 | A14 | A13 | B12 | C12 |
|------|------|------|------|------|------|
| mm. | mm. | mm. | mm. | mm. | mm. |
| 16.50 | 13.10 | 13.40 | 12.60 | 12.28 | 12.70 |
| 16.00 | 12.95 | 13.20 | 12.90 | 12.68 | 12.39 |
| 15.40 | 12.70 | 14.80 | 13.35 | 12.69 | 12.37 |
| 15.75 | 12.70 | 13.75 | 12.60 | 12.35 | 11.73 |
| 15.80 | 12.80 | 13.70 | 13.00 | 12.18 | 11.92 |
| 15.40 | 12.80 | 14.30 | 13.00 | 12.15 | 12.00 |

As the star grains become larger and larger the coating cannot be operated in one bowl, and they are divided into a few groups. At last they become a number of lots, each lot of stars being classified, packed and stored, and when the stars are loaded into a shell it is wise to fill the shell with stars taken from the same batch. When stars of a definite diameter are required, we must first select cores of the right size, and it is necessary to make frequent selections between the coating operations, though in practice the star grains are not selected in this way except in special circumstances.

Stars are manufactured mostly by hand nowadays in Japan. Some fireworkers use a coating machine, but there is a need in Japan to develop the demand for fireworks. For small quantities and large numbers of different types of stars the coating machine is not so adaptable. The machine must be so well cleaned when one composition changes to the next, that in the case of a star with layers of different colors or of manufacturing various kinds of stars, the machine operation is not as simple as the hand operation. The principle of manufacturing stars by machine is the same as that by hand, but the machine made stars are sometimes less easy to ignite because they tend to gain a higher specific gravity than those made by the hand method.

The manufacturing process with water insoluble binders is almost the same as the above process. In this case the use of large amounts of linseed oil make the stars unignitable. When a volatile solvent is used in place of water for the manufacture of stars care must be taken to keep the material away from flames, fire, electric heaters, hot electric lamps, and electric switches. Moreover it must be remembered that the vapors of organic solvents are often toxic.

*Firework Shells*

The firework shells are classified into two fundamental types, the so-called "Warimono" and "Poka". Warimono is the type which draws a chrysanthemum-like flower in the sky by driving light stars or smoke stars in all directions, and is also named "Chrysanthemum". Chrysanthemum shells have a large amount of bursting charge and a strong outside shell which helps the action of the bursting charge. The name "Poka" came from the weak sound of the explosion of the shells and is difficult to translate into English. Poka is not a Chrysanthemum type, for its purpose is only to throw out and spread the contents in the sky; it has a small amount of bursting charge, enough to break the shell into two parts along its joining line. It has a weak shell, which is strong enough not to be broken when it is fired. But in practice there are shells which lie between the two types. Most round shells of European type are almost the same as Poka in their effect.

*Warimono (Chrysanthemum)*

The construction of Warimono (Chrysanthemum) - the most representative construction of Warimono is shown as in Fig. 121. It can be seen that it consists of four main parts, i.e. shell A, star B, bursting charge C and fuse D. Shell A plays a role not only as the container for the stars and the bursting charge, but also of helping and adjusting the ejecting force of the bursting charge. The Stars B are ignited by the explosion of the bursting charge, C and fly like the petals of a chrysanthemum flower, burning on the surface. The bursting charge is ignited by the fuse D and breaks the shell into small pieces by the explosion which also ignites the stars and ejects them outwards in all directions. The fuse D is ignited on its outside end by the flame of the lifting charge in the mortar on shooting, and it keeps a delay time until the shell reaches a desired height in the sky and ignites the bursting charge just after the delay time has passed. The principal points of these constructions are described as follows:

Shell A consists of the outer shell a and inner shell b. Sometimes only the inner shell is called the "Shell", and in this case the outer

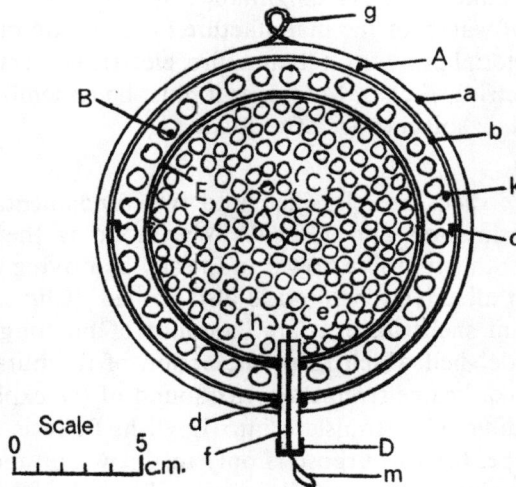

Fig. 121   Warimono shell

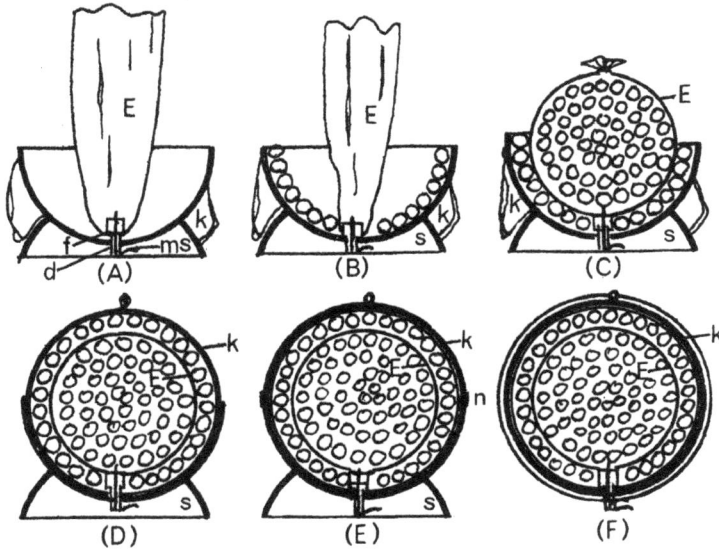

Fig. 122    Stages in shell manufacture

shell is called the "Pasted Shell". The inner shell b is only the container, and its strength is not very important. As the raw material of shell b, newspaper or cardboard paper is used and in Japan the newspaper shells are the most admired because of their uniform brittleness. The outer shell a is made of many folds of Japanese paper (Kozo) or of kraft paper, which is pasted on inner shell b to a thickness which is enough to give a suitable velocity to the stars on the explosion of the bursting charge C. The structures of the stars, bursting charge and fuse have been described above already. The process of assembly of a standard type of shell as in Fig. 121 is shown in order in Fig. 122. This process is applied to the shells of middle size e.g. 5 inches, 6 inches or 7 inches in diameter.

First the inner shell is prepared, a small hole being made in the center of the hemisphere of the shell, and a piece of fuse f is inserted through it, being fixed to the hemisphere with hemp d. The hemp is well pasted and well fixed on the shell by winding it around the fuse. This is a very important process to prevent fire from entering the shell on shooting. The outside end of the fuse is protected by being covered with a paper cap. On the cap(m) generally the name of the shell or the kind of the shell is registered. Before this fuse fixing operation a bag E made of Japanese paper, for the bursting charge, is attached to the inner end of the fuse, and the fuse is inserted into the hole of the hemisphere, from the inside. Then an outer bag k made of Japanese paper is pasted to the edge of the hemisphere, and it is turned over as in Fig. 122 (A). The bag k is used for the arrangement of the stars of the upper hemisphere. Two bags, E and k, are made generally of strong Japanese paper (Kozo) as described above, but recently we have found that they are often made of a rather weak copy paper. When we use a bursting charge of chlorate powder, the inner bag E must be made always of strong paper so that it cannot be broken during the arrangement of the stars. This avoids the bursting charge coming into contact with the stars and causing an accident, for the chlorate in the bursting charge together with the sulphur in the stars increases the sensitivity. The inner shell which is thus prepared is put on a rest S as in Fig. 122 (A). Fig. 122 (B) shows the arrangement of stars in the lower hemisphere. Then as shown in (C) an amount of the bursting charge is charged into the bag E and the hemisphere is tapped on the outside with a small wooden stick to avoid any looseness between the stars and the bursting charge. The bag is then sealed at the top being bound with a piece of hemp, the outer bag k is restored upwards and the stars of the upper hemisphere are arranged with the help of bag k as in (D). This arrangement is covered with another hemisphere on the top and it is pressed downwards by hand, being tapped with the wooden stick on the outside so that the two hemispheres meet each other at their edges. Sometimes the shell hemispheres have tally marks at the joining point of the two to make the operation easier. When the shell is correctly assembled the hemispherical shells meet each other with only a little pressure from the hand, thus avoiding any looseness between the stars and the bursting charge. If the hemispherical shells meet each other loosely, it shows that the amount of the bursting charge is too small, and it must be increased. The shell thus assembled has a ribbon tape pasted over the junction as in (E) shown by n, or it is fixed with another hemispherical shell, which is attached on the side of the assembled shell to prevent separation while the shell is sent to the next process, pasting. (F) shows the complete product.

The process described above is most adaptable for shells of middle size, but for smaller or larger shells the process must be managed conveniently according to the situation. The process shown in Fig. 123 is applied to a small shell of less than 4 inches in diameter. In this case

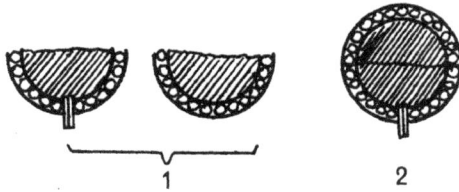

**Fig. 123** Filling small shells

the two hemispheres are prepared individually and then put together into one sphere.

Fig. 124 shows the process for larger shells of about 10—12 inches in diameter, and in this case (1) first the lower hemisphere is prepared and (2) then the upper hemisphere, in which a loading hole is made, is fixed to the lower hemisphere. Through this hole the bursting charge and the stars are loaded alternately, keeping the boundary between the bursting charge and the stars by the paper bag. (3) The hole is then sealed with the piece of shell which was previously removed from the upper hemisphere when the loading hole was made. During the process the sphere is tapped on the outside with a small wooden stick to assemble the contents tightly.

The processes described above are for the Chrysanthemums with single petalled flowers. The processes for double petalled flowers or special multipetalled flowers are described as follows: Fig. 125 shows the simplest form, and in this case the inner petal stars are mixed in the bursting charge. Fig. 126 shows the bursting charge composition pasted on the inner petal stars. Fig. 127 shows a more elaborate shell where a part of the bursting charge is first shaped into a sphere by being wrapped with thin paper and wound on the outside with string.

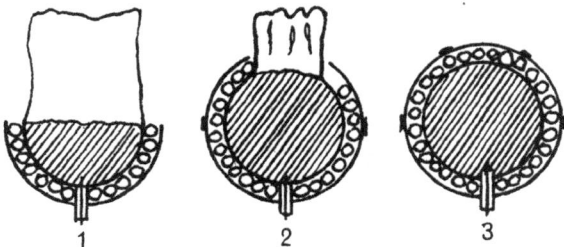

**Fig. 124** Filling large shells

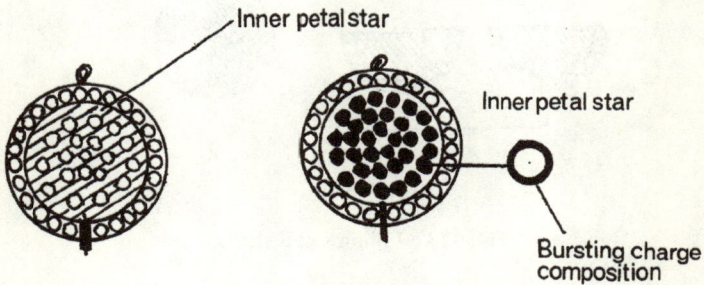

**Fig. 125,126**   Shells with inner petal stars

The inner petal stars are then arranged outside the bursting charge sphere and wrapped again on the outside by the same process. This is the inner ball, which creates inner petals. Putting this ball at the center of the shell, we assemble and complete the shell by much the same way as described above.

Fig. 128 shows the process for a multipetalled Chrysanthemum and it is assembled in the way described above, the only difference being to first make the inner double petalled ball. Fig. 129 shows the process for another elaborate inner multipetalled ball, i.e. a paper tube, which becomes the axis of the ball, and perforated at the center. The bursting charge and petal stars are arranged around the tube alternately, so that the hole is at the center of the ball. On assembly one end of the tube is connected to the fuse. This kind of ball with an axis is called "Nuki-shin".

The stars which are placed at the center, do not fly so far when the

**Fig. 127**   More complex inner petal star shell

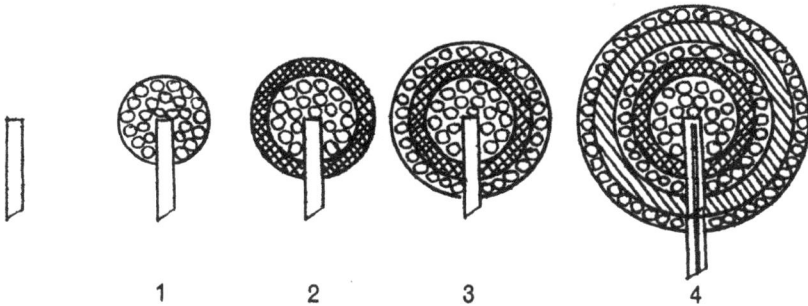

Fig. 128   Assembly of multi-petalled shell

exploding charge bursts, and the duration of the inner stars must be made shorter than the outer stars. Moreover the grain size of the inner bursting charge must be smaller than that of the outside stars to strengthen the action of the bursting charge. If it is not planned in this way, we do not get a good flower.

The pasting process for making the outer shell is as follows:

As Fig. 130 shows, long pieces of paper are pasted on the inner shell like the stripes on a watermelon. In this case A-A' is taken as an axis, and the points A and A' become the poles. The procedure is as follows: First a paper tape is pasted around the assembled inner shell along the seam of the two hemispheres to prevent them from separating. First the axis is taken so that it becomes perpendicular to the seam. The paper pieces are made by cutting a sheet of paper so that the direction of the fibers is at right angles to the length of the paper pieces, in order to fit closely around the shell. Japanese paper, which is previously

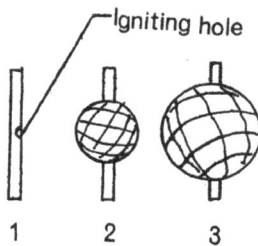

Fig. 129   Shell with an axis

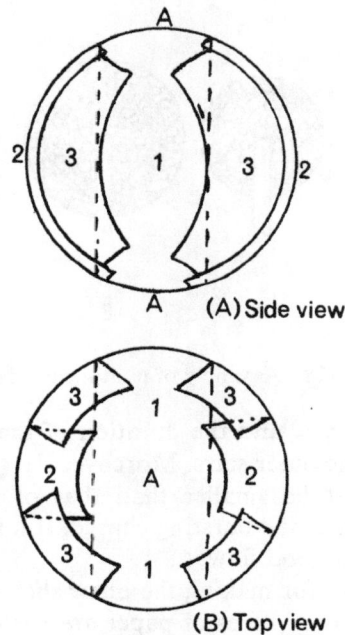

**Fig. 130**   Making outer shell covers

pasted in two or three folds is normally used. Kraft paper is so cheap
and strong, that it is also used as the outer shell material. The paper
pieces are prepared as follows: The paste, is made of one part of wheat
flour and three parts of water, but some prefer to use a paste made of
nonglutinous rice. The paste is spread on one side of the paper with a
wooden spatula, and the pasted surface is folded in half, the pasted
sides together. The purpose of this is to protect the paste from drying
during the pasting operation of the outer shell. The pasted paper is cut
into long pieces keeping them folded. The pasted shell with one layer is
shown in Fig. 130. The width of the paper pieces must be constant as
far as possible and the folding of each paper piece with another piece
on the shell should be avoided if possible, as it is shown in (B), so that
the shell may be uniform. In Fig. 130 the figure on each piece of paper
pasted on the shell shows the order of the pasting. Then two other
poles are taken so that the new axis is placed at right angles to the

axis A—A'. and the pasting process is repeated in the same way as before. When two or three layers of paper have been added, a process called "Gorokake" is applied; the pasted shell is placed on a thick wooden plate and another small wooden plate is placed on the shell, which is then pressed and rolled between the two plates. The edges of the pieces of paper on the shell are then perfectly fixed on the shell which becomes a perfectly round sphere. After this process is finished, the shell is dried in the sun. When it is well dried, the same process (i.e. pasting, Gorokake and drying) is repeated until the paper folds (i.e. the thickness of the layer of pasted paper) reaches the required degree. The necessary number of layers of the pasted paper is proportional to the diameter of the shell, according to the experience of fireworkers, and in ordinary cases, with Japanese Kozo paper 5 folds per 1 inch or 7 folds per 1 inch in diameter is recommended. For example in the case of a 4 inch shell (in diameter) the number of all folds is $4 \times 5 = 20$, and in the case of a 6 inch shell (in diameter) it is $6 \times 5 = 30$, when we take the 5 folds per inch. These two shells have the same breaking strength against the bursting pressure according to the experience of fireworkers, and this is proved by the calculation based on the stress strain theory. Namely the folds of the pasted paper must be increased proportionally with the shell size to keep the same breaking strength. Fig. 131 shows a section of a shell which is pasted in this way. In the case of kraft paper the strength is twice that of Japanese "Kozo" paper, and the necessary layers become half of the above values.

a. Kozo paper, 7 folds per inch.

b. Kozo paper, 14 folds per inch.

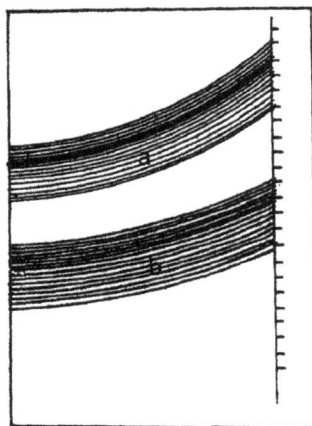

Photo 6   Sections of Shells
(6 inch shell in dia.)

**Fig. 131**   Sections of shell

*Poka*

*The construction of Poka.* The construction of the shell "Poka" is quite different from that of the shell "Warimono", i.e. Chrysanthemum. The shell has chambers in which various air floating objects are packed. The principle of the arrangement of the chambers in the shell is shown in Fig. 132. A is the simplest one, and consists of one fire chamber, but B has two chambers, a fire chamber and a fire prevention chamber. The fire chamber is so constructed that the flame of the bursting charge easily touches the contents. The fire chamber contains stars, reports, small flowers, falling leaves, comets, whistles etc. which must be ignited without fail. The fire prevention chamber is constructed in such a way that a wall is placed between the two chambers to protect the contents from the fire of the bursting charge when the shell explodes. This "wall" consists of cardboard or cotton seeds. The wall also acts as stuffing to prevent the contents from moving.

The method of fixing the fuse to the shell is the same as in the Warimono shells.

The amount of the bursting charge for Poka should be enough to break the shell into two pieces (i.e. the hemispheres), and it is fixed in the shell as shown in Fig. 132 and Fig. 133. The method in Fig. 132 is to fix the charge on the inside end of the fuse, and is rather troublesome, but the bursting charge is well ignited by the fuse without failure.

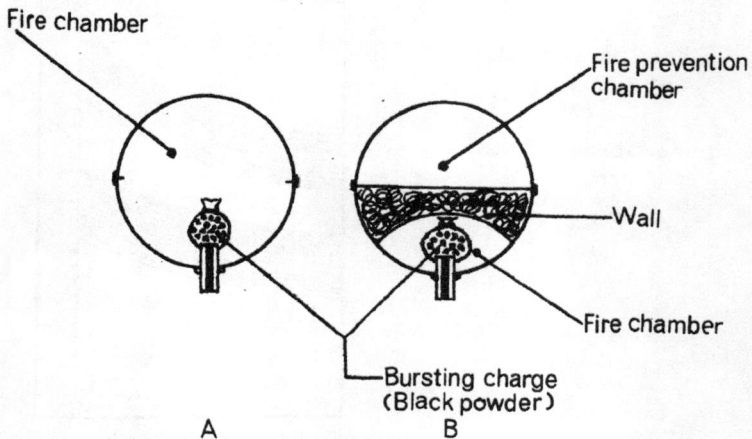

**Fig. 132**  Poka shell construction

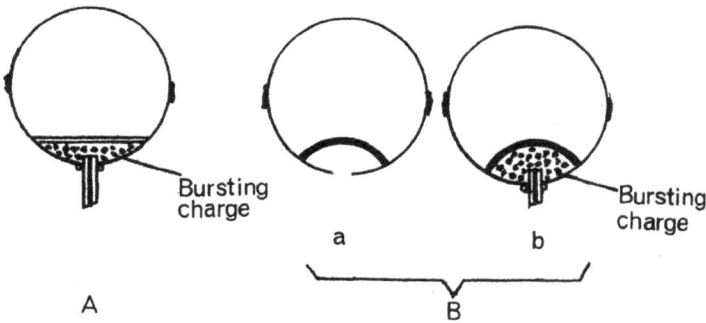

**Fig. 133**　Poka shell construction

Fig. 133 A shows a more simple method. In this case a few pieces of "powder-pasted" paper are inserted in the charge for perfect ignition. Fig. 133 B shows the shell with no bursting charge as in a, which is afterwards charged and fixed with a fuse as in b, when it is used. It can be handled as "paper" without the bursting charge during storage and transportation, and it is very convenient when it is used for paper Balloons called "Fukuromono" and Flags.

Fig. 134 shows the constructions of representative Poka shells. A is a report shell, "Thunder", which makes three or five reports with the same time intervals. B is a Flag or Balloon and C is a Flare, in which the fire prevention wall is made of cotton seeds to protect the parachute from the fire. The construction of Chains are the same as that of Flares.

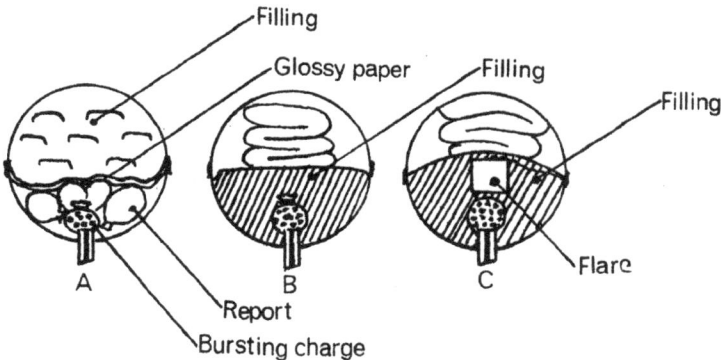

**Fig. 134**　Poka shells with flash reports or paper figures

*Planning of Poka*   Here the relation between the strength of the shell and the amount of the bursting charge must be considered. The shell must be strong enough not to be broken when it is fired, and if the weight of the contents is large, we must think also of the possibility of "Head Breaking", which means that the shell is broken by the shock of the firing and the contents are immediately projected out of the shell. It is necessary to maintain the strength of the shell during firing, but it is wise to have the minimum amount of the bursting charge, enough to break the shell into two hemispheres. The folds of the paper pasted on the shell are just adequate for these conditions. A few examples of the amount of the bursting charge are shown as follows:

*Table 8.*   The Bursting Charge of Poka
(In the case of black powder)

| Diameter of Shell (inches) | Amount of Charge (grams) | Diameter of Shell (inches) | Amount of Charge (grams) |
|---|---|---|---|
| 3 | 2.0 | 7 | 6.0 |
| 4 | 3.0 | 8 | 7.0 |
| 5 | 4.0 | 10 | 8.0 |
| 6 | 5.0 | 12 | 10.0 |

The process for pasting paper on the assembled shells is the same as that of Chrysanthemum. The folds of the paper layer of the outer shell is 1.7—2.2 per inch in diameter, on the seam of the two hemispheres and the average value is about 2.1 on the same part. The weakest position of the completed shell is the junction of the two hemispheres, and when the shell explodes, it is broken along this line. The inner shell is made of newspaper, and as described above its strength is unimportant in the case of the Chrysanthemum, but is important in the case of Poka, because the ratio of the strength of the inner shell to the whole is higher than that of the Chrysanthemum. 3mm. in thickness of a newspaper shell corresponds to the eight folds of Japanese Kozo paper in strength. Accordingly (e.g. in the case of 4 inch shell - in diameter -) the paper pieces are first pasted on and along the seam two or three times, and then they are pasted on the whole surface of the sphere just once. When the strength of the pasted paper along the junction is too great, the shell cannot be broken along the line, but is perforated in some other part by the explosion gas, which escapes out of the shell through this hole, but does not project the contents out of the shell.

# References

1. A.St.H.Brock, Pyrotechnics, O'Connor (1922).
2. A.St.H.Brock, A History of Fireworks, Harrap (1949).
3. J.R.Partington, A History of Greek Fire and Gunpowder, Heffer (1960).
4. A.A.Shidlovsky, Osnovy Pirotekhniki (Fundamentals of Pyrotechnics) 1st ed. Moscow (1943).
5. H.E.Ellern, Modern Pyrotechnics, Chemical Publishing (1961). Military and Civilian Pyrotechnics (1968).
6. G.W.Weingart, Pyrotechnics, Chemical Publishing 1st ed. (1947)
7. T.L.Davis, Chemistry of Powder and Explosives, Wiley (1941).
8. A.Izzo, Pirotecnia e Fuochi Artificiali, Hoepli, Milan (1950).
9. H.B.Faber, Military Pyrotechnics, Government Printing Office Washington (1919).
10. F.Ullmann, Encyklopadie der Technischen Chemie, Munchen (1963).
11. Kirk-Othmer, Encyclopedia of Chemical Technology, Interscience Publishers, New York.
12. British Intelligence Objectives Sub-Committee Reports, Numbers 461, 477, 1233, 1313. Reports of the Combined Intelligence Objectives Sub-Committee.
13. S.Fordham, High Explosives and Propellants, Pergamon, (1966).
14. D.B.Chidsey, Goodbye to Gunpowder, Alvin Redman, (1964).
15. Edwards & Wray, Aluminum Paint and Powder, Aluminum Co of America
16. Shellac, Angelo Rhodes Ltd.
17. Guide to the Explosives Acts, Her Majesty's Stationery Office.
18. N.Heaton, Outlines of Paint Technology, Griffin.
19. T.Kentish, The Pyrotechnists Treasury, Chatto & Windus (1887).
20. W.H.Browne, The Art of Pyrotechny, The Bazaar c. (1880).
21. H.G.Tanner, Instability of Sulfur-Potassium Chorate Mixture. Journal of Chemical Education Vol. 63, No.2, Feb. 1959.
22. Watkins, Cackett & Hall, Chemical Warfare, Pyrotechnics and the Firework Industry, Pergamon, (1968).
23. T.Shimizu, Hanabi (Fireworks)[in Japanese] Tokyo (1957).

# Index

www.ingramcontent.com/pod-product-compliance
Lightning Source LLC
Chambersburg PA
CBHW021429180326
41458CB00001B/193